Geographic Perspectives on Urban Sustainability

The 21st century has been called the "century of the city." Unprecedented and uneven urban growth and expansion coupled with climate change have compounded concerns that current urbanization pathways are not sustainable. Calls for scholarship on urban sustainability among geographers cite strengths in both examining human-environment interactions and unravelling urbanization patterns and processes that positioned the discipline to make unique contributions to critical research needs.

Geographic Perspectives on Urban Sustainability reflects on the contributions that geographers have made to urban sustainability scholarship on varied domains such as transportation, green infrastructure, and gentrification. Contributed chapters probe uniquely geographic perspectives on urban resilience, environmental justice, political ecology, and planning that arise from empirically integrating social and biophysical realms that arise from considering spatial dimensions of problems like scale- and place-based peculiarities of phenomena.

This book will be of great value to scholars, students, and policymakers interested in Urban and City Planning, Political Ecology, and Sustainable Urbanism.

The chapters in this book were originally published as a special issue of *Urban Geography*.

V. Kelly Turner is Assistant Professor of Urban Planning and Associate Director of Urban Environmental Research at the Luskin Center for Innovation in the Luskin School of Public Affairs at the University of California, USA. She holds a PhD in Geography from Arizona State University, Tempe, USA.

David H. Kaplan is Professor of Geography at Kent State University, USA. He is the past President of the American Association of Geographers and the Editor-in-Chief of *Geographical Review*.

Geographic Perspectives on Urban Sustainability

Edited by
V. Kelly Turner and David H. Kaplan

LONDON AND NEW YORK

First published 2021
by Routledge
2 Park Square, Milton Park, Abingdon, Oxon, OX14 4RN

and by Routledge
52 Vanderbilt Avenue, New York, NY 10017

Routledge is an imprint of the Taylor & Francis Group, an informa business

© 2021 Taylor & Francis

All rights reserved. No part of this book may be reprinted or reproduced or utilised in any form or by any electronic, mechanical, or other means, now known or hereafter invented, including photocopying and recording, or in any information storage or retrieval system, without permission in writing from the publishers.

Trademark notice: Product or corporate names may be trademarks or registered trademarks, and are used only for identification and explanation without intent to infringe.

British Library Cataloguing-in-Publication Data
A catalogue record for this book is available from the British Library

ISBN13: 978-0-367-67193-8

Typeset in Minion Pro
by codeMantra

Publisher's Note
The publisher accepts responsibility for any inconsistencies that may have arisen during the conversion of this book from journal articles to book chapters, namely the inclusion of journal terminology.

Disclaimer
Every effort has been made to contact copyright holders for their permission to reprint material in this book. The publishers would be grateful to hear from any copyright holder who is not here acknowledged and will undertake to rectify any errors or omissions in future editions of this book.

Contents

Citation Information vi
Notes on Contributors vii

1 Geographic perspectives on urban sustainability: past, current, and future research trajectories 1
 V. Kelly Turner and David H. Kaplan

2 Transportation sustainability in the urban context: a comprehensive review 13
 Selima Sultana, Deborah Salon and Michael Kuby

3 Urban resilience for whom, what, when, where, and why? 43
 Sara Meerow and Joshua P. Newell

4 Green infrastructure, green space, and sustainable urbanism: geography's important role 64
 Lisa Benton-Short, Melissa Keeley and Jennifer Rowland

5 Uneven urban metabolisms: toward an integrative (ex)urban political ecology of sustainability in and around the city 86
 Innisfree McKinnon, Patrick T Hurley, Colleen C Myles, Megan Maccaroni and Trina Filan

Index 113

Citation Information

The chapters in this book were originally published in the *Urban Geography*, volume 40, issue 3 (2019). When citing this material, please use the original page numbering for each article, as follows:

Chapter 1
Geographic perspectives on urban sustainability: past, current, and future research trajectories
V. Kelly Turner and David H. Kaplan
Urban Geography, volume 40, issue 3 (2019) pp. 267–278

Chapter 2
Transportation sustainability in the urban context: a comprehensive review
Selima Sultana, Deborah Salon and Michael Kuby
Urban Geography, volume 40, issue 3 (2019) pp. 279–308

Chapter 3
Urban resilience for whom, what, when, where, and why?
Sara Meerow and Joshua P. Newell
Urban Geography, volume 40, issue 3 (2019) pp. 309–329

Chapter 4
Green infrastructure, green space, and sustainable urbanism: geography's important role
Lisa Benton-Short, Melissa Keeley and Jennifer Rowland
Urban Geography, volume 40, issue 3 (2019) pp. 330–351

Chapter 5
Uneven urban metabolisms: toward an integrative (ex)urban political ecology of sustainability in and around the city
Innisfree McKinnon, Patrick T Hurley, Colleen C Myles, Megan Maccaroni and Trina Filan
Urban Geography, volume 40, issue 3 (2019) pp. 352–377

For any permission-related enquiries please visit:
http://www.tandfonline.com/page/help/permissions

Contributors

Lisa Benton-Short Department of Geography, George Washington University, Washington, D.C., USA.

Trina Filan Independent Scholar, Helena, USA.

Patrick T Hurley Environmental Science, Ursinus College, USA.

David H. Kaplan Department of Geography, Kent State University, USA.

Melissa Keeley Department of Geography, George Washington University, Washington, D.C., USA.

Michael Kuby School of Geographical Sciences and Urban Planning, Arizona State University, Tempe, USA.

Megan Maccaroni Environmental Science, Ursinus College, USA.

Innisfree McKinnon Social Science Department, University of Wisconsin Stout, Menomonie, USA.

Sara Meerow School of Natural Resources and Environment, University of Michigan, Ann Arbor, USA.

Colleen C Myles Department of Geography, Texas State University, San Marcos, USA.

Joshua P. Newell School of Natural Resources and Environment, University of Michigan, Ann Arbor, USA.

Jennifer Rowland Research and Advocacy Associate, Public Lands Project at Center for American Progress, Washington, D.C., USA.

Deborah Salon School of Geographical Sciences and Urban Planning, Arizona State University, Tempe, USA.

Selima Sultana Department of Geography, The University of North Carolina at Greensboro, USA.

V. Kelly Turner Department of Geography, Kent State University, USA.

Geographic perspectives on urban sustainability: past, current, and future research trajectories

V. Kelly Turner and David H. Kaplan

ABSTRACT
The 21st century has been called the "century of the city" and compounded concerns that current development pathways were not sustainable. Calls for scholarship on urban sustainability among geographers cites strengths in the human-environment and urban subfields that positioned the discipline to make unique contributions to critical research needs. This special issue reflects on the contributions that geographers have made to urban sustainability scholarship. We observe that that integration across human-environment and urban subfields reflects broader bifurcations between social theory and spatial science traditions in geography. Piggy-backing on the rise of sustainability science, the emergence of urbanization science compels geographers to reflect upon the ways in which we are positioned to make unique contributions to those fields. We argue that those contributions should embrace systems thinking, empirically connect social constructs to biophysical patterns and processes, and use the city as a laboratory to generate new theories.

Introduction

Urban sustainability scholarship as an outgrowth of sustainability science has been established as a major interdisciplinary academic research priority, with recent calls for a new field of "urbanization science" that would generalize our understanding of human and environmental processes shaping cities (Alberti, 2017; Seto, Golden, Alberti, & Turner, 2017; Seto & Reenberg, 2014; Solecki et al., 2013; Wigginton, Fahrenkamp-Uppenbrink, Wible, & Malakoff, 2016). Geographers have previously commented that the discipline *appears* to be well positioned to contribute to urban sustainability scholarship due to long traditions in both human-environment and urban subfields and interest in interdisciplinary integration (Baerwald 2010; Bettencourt & Kaur, 2011; NRC, 1999 ; NSF 1999; Hanson & Lake, 2000; Kates, 2011; Mooney et al., 2013; Robbins, 2007; Skole, 2004; Turner, 2002; Wolch, 2007; Zimmerer, 2001). This special issue is motivated by the observation that integration across human-environment and urban subfields reflects broader bifurcations between social theory and spatial science traditions in geography. Specifically, urban sustainability scholarship within the urban geography subfield and political ecology within the human-environment subfield look to social theory to understand environmental themes in cities, but rarely integrate environmental data on biophysical dynamics into empirical studies (e.g., Longley, 2002; Walker, 2005). Urban sustainability scholarship in the land systems science tradition within human-environment geography integrates environmental data into spatial science frameworks to understand coupled human-biophysical dynamics in urban regions, but rarely integrates existing explanatory theory from the social sciences (Munroe, McSweeney, Olson, & Mansfield, 2014). Moreover, geographers aligned with land system science have already led development of interdisciplinary sustainability and urbanization science fields, while the contributions of geographic traditions aligned with social theory are less apparent in those realms.

In this introduction, we review the intellectual, methodological, and institutional dynamics in human-environment and urban geography that shape integrative urban sustainability scholarship in the field. We argue that some lines of urban sustainability scholarship in geography may miss an opportunity to shape the field of urbanization science if the level of integration between urban and human-environment subfields in geography remains siloed between social theory and spatial science. Importantly, we offer avenues to advance a more holistic integrative approach that dovetail with the urbanization science approach and introduce examples that highlight those approaches in this special feature.

Urban themes in human-environment geography

Human-environment geography has major scholarly traditions in land systems science and political ecology, informed by spatial science and social theory, respectively (Turner & Robbins, 2008; Zimmerer, 2010). Both traditions emphasize topics such as land management and livelihoods in resource dependent communities, human induced land degradation and conservation of natural resources, and human and environmental vulnerability because of development (Turner & Robbins, 2008). These areas of common topical interest have historically steered research towards non-urban systems.

Remote sensing data have been central to land systems science approaches and, until recently, been too course to provide meaningful insights into urban land dynamics (Wentz, Seto, Myint, Netzband, & Fragkias, 2011; Yang, 2011). Foundational work in land systems science, therefore, tended to frame urbanization as one of many drivers of land consumption and degradation and did not address intra urban dynamics (Turner, Lambin, & Reenberg, 2007). Recent insights into phenomena like the urban climate have accelerated more nuanced scholarship on cities as complex land systems (Georgescu, Morefield, Beirwagen, & Weaver, 2012; Stokes & Seto, 2016, Wentz et al., 2011). Urban land systems scholarship is leading development of fields like urbanization science because the spatial science approach favored in land systems science corresponds well to environmental sciences, seeks generalizable knowledge of urban systems, and considers how best to corral that knowledge into actionable insights to effect policy change (Magliocca et al., 2018; Zimmerer, 2010). Yet, land systems science struggles at times to empirically link social and ecological causes and consequences, especially those observable at local scales (Rindfuss, Walsh, Turner, Fox, & Mishra,

2004). As a result, land systems science empiricism has held traction in global forums like IPCC, but contributed less to local decision-making in cities.

Political ecology began to shift toward urban themes earlier than did land systems science. The subfield is less constrained by remote sensing and GIS technology and has been "suspicious" of the politics of these research tools (Turner & Robbins, 2008). In fact, analytic approaches, like metabolism studies of flows between material resources and capital are well suited to urban analysis (Swyngedouw & Heynen, 2004). Conventional themes in political ecology – social production of nature and environmental justice – bring cities into the foreground. The emerging subfield of urban political ecology challenges the logic of spatial patterns observed through land systems science approaches (Robbins, 2007; Swyngedouw & Heynen, 2004). Urban political ecology also aligns with urban geography perspectives that draw from social theory, as is evidenced by several publications in this journal (Cooke & Lewis, 2010; Keil, 2003, 2013; Myers, 2008; Quastel, 2009, Quastel, Moos, & Lynch, 2012; Ranganathan & Balazs, 2015). Yet, the political ecology tradition often fails to fully capture nature in practice, emphasizes representation over generalization, and it avoids policy prescriptions, which impedes collaboration with the natural sciences (Blaikie, 2012; Walker, 2005). At best, integrating qualitative case study and environmental data sets presents methodological and epistemological challenges and, at worst, some in the environmental science community view qualitative case study data as less rigorous (Meyfroidt, 2016, Petts, Owens, & Bulkeley, 2008).

The biophysical environment in urban geography scholarship

Historically, urban geography developed in tandem with urban sociology for which the natural environment was not at the forefront of scholarship concerning the place of the city within an urban system (e.g., Berry & Garrison, 1958; Getis & Getis, 1966;). Despite the metaphorical use of the term "ecology" of the city (Park, Burgess, & McKenzie, 1984), nature has remained absent later quantitative assessments of human activities and how they shaped urban space within and between cities (Abler, Adams, & Gould, 1971; Kaplan & Holloway, 2014; Pred, 1964; Wehrwein, 1942; Wheeler, 1993). Structuralist and humanist paradigms entered urban geographic explorations of cities in the 1970 and 1980s, but these perspectives largely focused on either political economic structures or human agency and perception (Harvey, 1973; Ley, 1983). The Los Angeles School that emerged as a competing approach to the Chicago School in the 1990 and 2000s may come closest to engaging the role of the natural environment in cities (Dear, 2002a, 2002b; Nicholls, 2010; Shearmur, 2008). Dominated by Marxist and poststructuralist social theorists, however, this line of scholarship was primarily concerned with the politics of nature and far less attentive to biophysical processes. Ashmore and Dodson (2017, p. 105) observe that "it is curious that urban geography as a specialization has been seen within geography as tacitly the preserve of human geographers." When Wheeler (1993) examined the articles published in *Urban Geography* between 1980 and 1991, none of the 18 themes that he identified related to nature and the environment. Even more contemporary "readers" reexamined urban geography, but included very little about urban environmental or sustainability concerns (Fyfe & Kenny, 2005). Urban space and place, at least, was "devoid" of nature (Hanson, 1999).

There was a missed opportunity to engage the work of physical geographers in the 1970s and 1980s (Berry & Horton, 1974; Detwyler & Marcus, 1972; Douglas, 1981, 1983). They argued that the city was a particular kind of ecosystem, with feedback loops between human and biophysical system components (similar to the aforementioned urban land systems work), and that any search for answers to the social ills of cities must include an understanding of the biophysical dimensions of urban issues alongside the socio-cultural explanations (Douglas, 1981). Dynamics within geography at the time potentially explain this lost opportunity for collaboration between physical and urban geographers. The cultural turn within urban geography was ill fit with the systems approaches favored by physical geographers. Physical geographers were reluctant to interface with social theory, especially cultural ecology framings, based on their fears of environmental deterministic explanations. Whatever the reason for geographer's reticence to contribute, an ecology of the city that empirically integrated ecological and social realms did develop in the ensuing years (Grimm et al., 2008). The absence of social theorists in this new urban ecology creates explanatory holes and, sometimes, misappropriation of social science constructs (Grove et al., 2015; Ramalho & Hobbs, 2012).

The concept of sustainability, harnessed to urban design, planning, and economic development, have invigorated urban geography research on environmental themes (NSF, 1999, Wolch, 2007). Design-oriented perspectives seek to mitigate the social and environmental ills associated with urban sprawl through smart growth, transit-oriented development, conservation design, and New Urbanism among others (Arendnt et al., 1996; Duany, Plater-Zyberk, & Speck, 2000; McHarg, 1969; Reid & Cervero, 2001). Some of the design goals such as building to a human scale, increasing walkability, and embracing socio-economic diverse communities resonated among urban geographers (Ellis, 2002; Platt, 2006; Talen, 2002). For the most part, design-oriented approaches have drawn sharp criticism among urban geographers that have argued that much of what is described as sustainable design is environmental design and ignores the social processes (Braun, 2005; Talen, 2002). For instance, this journal has featured multiple articles on the New Urbanist design school, including a special feature in 2001. These articles argue that New Urbanism at best produces mixed social and environmental outcomes or at worst constitutes a green marketing tool that, perversely, enables gentrification and sprawl (Al-Hindi & Till, 2001; Till, 2001; Trudeau & Kaplan, 2016; Trudeau & Malloy, 2011; Zimmerman, 2001). Given its emphasis on social theory, it is unsurprising that urban geography scholarship engaging with environmental design theories has produced little in the way of empirical evidence of environmental outcomes.

Entrenching divides via academic institutions

Academic institutions have reinforced the division between the human-environment and urban geography subfields, especially by pulling geographers from the land systems science traditions into interdisciplinary society-environment fields. Much of the geographic work on human-environment interactions has come under the umbrella of sustainability science, an interdisciplinary field recognized by top scientific venues such as *Proceedings of the National Academy of Sciences* and *Science*, major funding bodies like the National Science Foundation (NSF) Coupled Natural Human Systems Program, and international research agendas and consortiums such as the *International Geosphere-Biosphere Programme* (IGBP, http://www.igbp.net/), *Future Earth* (http://www.futureearth.org/), and the *Intergovernmental Panel on Climate Change* (http://www.ipcc.ch/) (Cash et al., 2003; Clark, 2000; Clark, 2007; Clark & Dickson, 2003; Lubchenco, 1998; Kates et al., 2000). Land System Science in particular has formal institutional linkages to the *Global Land Programme* (GLP, https://glp.earth/) within the IGBP: large, international, interdisciplinary science consortiums (Meyfroidt, 2016; Seto & Reenberg, 2014; Seto et al., 2012; Seto, Sánchez-Rodríguez, & Fragkais, 2010). Concurrently, ecology and landscape ecology, fields that historically viewed human actions as "disturbances" to the natural environment, began to embrace humans as integral to ecosystems and urban regions as ecosystems in their own right (MEA date, Crutzen, 2002; Forman, 2014; Grimm et al., 2008). Converging intellectual streams between sustainability science, ecosystem ecology, and other nature-society fields (Miller et al., 2008) have become institutionalized now with the emergence of *Future Earth* (http://www.futureearth.org/), an umbrella consortium for the various international scientific associations across these fields. In short, sustainability science has become big science.

While the interdisciplinary sustainability science field has consolidated in several respects, the interdisciplinary field of urban studies is far more fragmented. Urban Studies was established in the 1960s to examine urbanization and the processes that shape cities and urban life. Urban geographers are active in urban studies, contributing their time in editing urban studies journals and participating in urban studies organizations, but Urban Studies has not developed a singular theoretical framework or the institutional infrastructure that precipitated growth in interdisciplinary environment-society fields. Urban Studies as a whole is made up of many research traditions and has no clear "big science" analog to sustainability science. What is more, few universities outside of North America have urban studies programs and scholars that would populate such a program have remained in the social sciences or architecture (Hutchison, 2010). So there is little reason for out-migration of the kind seen in interdisciplinary environment-society fields.

Work on urban sustainability within sustainability science and urban studies communities remains scattered, much like the scholarship within geography itself (Alberti, 2017; Solecki et al., 2013). Many within this community yearn for a more unified set of approaches, tools, and theories to the study of cities. A recent publication written by three geographers calls for an "urbanization science" – the study of urbanization processes and how they intersect with environmental systems. It argues that coalescence around a set of urban themes and approaches is hampered by the fragmentations that beset urban studies (Solecki et al., 2013). Yet evidence suggests that the emerging urbanization science community is intentionally mimicking the organization of the sustainability science community (Frank & Gabler, 2006). At the same time, the sustainability science community is increasingly concerned with urban themes as evidenced by the *Urbanization and Global Environmental Change* (https://ugec.org/) program (2006--2017), recent special features in *The Proceedings of the National Academy of Sciences* and *Science* on urban sustainability, and a new journal for *Urban Science* (Alberti, 2017; Seto et al., 2017; Wigginton et al., 2016). The near future will likely see the development of a cohesive, big science community, unified by systems-based frameworks on urban sustainability themes, and this could include a number of geographers.

Advancing integrative geographical research on urban sustainability

If a scholarly community on urban sustainability is developing at the interdisciplinary "borderlands" with geography (Zimmerer, 2010), we might ask what contributions are uniquely geographic and how might greater collaboration between human-environment and urban subfields improve such understanding. As Robbins (2007, p. 313) reminds us, our discipline is strongest when it embraces synthesis to reveal "counterintuitive findings that turn problems on their heads, compel new avenues for thinking, and stress the emergent, dialectical, or downright strange lessons of geography." The articles in this feature suggest that a set of uniquely geographic, counterintuitive insights that will find greater traction if they (1) embrace systems thinking, (2) empirically connect biophysical patterns and processes with social constructs, and (3) use the city as a laboratory to generate general new theories of human-environment interactions.

Embrace systems thinking

Systems thinking means that individual phenomena are understood in relation to multiple social and environmental factors. Yet, much of urban sustainability literature has focused on constituent parts of the city – transportation, parks, housing, and so forth. In their review of transportation sustainability research, Sultana et al argue for greater connections between transportation and land use scholarship in which transit systems are characterized as both the cause and outcome of urban land systems. Systems thinking involves making multi or – interdisciplinary connections. In their review of municipal sustainability plans, Keely et al. show that such plans would be best informed by scholarship that can interrogate the interrelationships between urban domains. Geographers consistently aver how scale is significant to framing problems, yet urban sustainability scholarship often suffers from domain specific, scale-dependent research areas (Keely et al., McKinnon et al.). Systems in an urban context requires consideration of the relationship between cities, their hinterlands, and even distal urban locations. For instance, in their synthesis of urban and exurban political ecology theory, McKinnon et al. show that urban political ecology privileges macro-scale issues around decision-making while exurban political ecology views local-scale conflicts as stemming from the social construction of narratives about nature. They call for expanding notions of the city beyond the urban core, enabling integration of macro- and local-scale social processes. Finally, the physical sciences often dominate systems thinking but struggle to conceptualize the human components of social-ecological systems. However, a geographic perspective can move beyond critique and offer points of departure for looking under-the-hood of the human systems' black box. Meerow and Newell offer an empirical approach for integrating fundamental questions about the social goals of resilience-based land management into empirical methods.

Empirically connecting social constructs to biophysical patterns and processes

A potential strength in geography for advancing urban sustainability research is the coexistence of social and biophysical approaches; yet, more work is necessary to fully leverage potential integration across those domains that could generate theories that empirically connect social and environmental realms. For sustainable transportation research, Sultana et al comment that early assessments of transportation used concepts such as carrying capacity and IPAT to develop the ASIF transportation framework that relates behavior to biophysical variables (mode share, energy, and carbon content). More recent work on transportation sustainability has become more siloed, with scholarship focusing on transit as either a techno-engineering problem or behavior, equity, and health issue, and empirical integration is needed in future scholarship. Establishing empirical connections between social and biophysical data creates a number of challenges, many of which geographers are well positioned to address. For instance, Keely et al. highlight how the selection and analysis of biophysical data (e.g., what constitutes urban green space) implicitly biases the results of studies that use spatial data to examine distribution and access to green space. Similarly, Meerow and Newell demonstrate how urban green space planning can provide a pluralistic understanding of urban resilience through comparison of multiple sets of social goals using spatial models. They show that different social goals direct green space to different locations within the city. Moreover, they determine that the scale of analysis influences what goals get emphasized in the first place. Social theory especially holds promise in providing explanations for the patterns and processes observed in spatial models of land use change. For instance, McKinnon et al. show that patterns of exurban development are tied to macro-scale causal factors such as the economic power of the urban middle class and the development industry as well as local-scale causal factors such as the construction of the urban fringe as site of "escape" from urban life. McKinnon et al. insights would be especially salient for explaining the processes that drive urban land "teleconnections" – the social processes that connect urban sites to proximate and distal locations and the associated land patterns – a framework put forth by the urbanization science community (Seto et al., 2012).

Using the city as a laboratory to generate new theories

Geographers have a long tradition of using cities as field sites to collect empirical data. Advancing an urban sustainability research agenda should involve doubling down on this notion of the-city-as-laboratory, and framing policies, build infrastructure, or other urban interventions as "experiments" in urban sustainability. For instance, Sultana et al argue that transportation research should frame policy change – and, specifically, policy packages – to tease out which factors produce differences in travel behavior, equity, and adoption of more sustainable modes of transportation. Moving beyond correlation will require longitudinal data sets or, at a minimum, before-and-after data on variables of interest such as land use and travel behavior. Indeed, while policy and design interventions are often empirically informed, the actual impact of such interventions in real-world settings are often not monitored. Keely et al. observation that both the types and benefits of urban green space are heterogeneous reveals how municipal sustainability plans bundle a range of potential benefits to justify specific green infrastructure projects. Geographers can determine what specific benefits are achieved from green infrastructure interventions and where do those projects balance environmental, economic, and social benefits in practice. Similarly, Meerow and Newell emphasize how real world outcomes of planning reveal relationships between interpretations of resilience and the spatial patterns they produce. These would add field-based data to theory driven models, like the scenarios produced in their study, and likely shed light on the social processes that drive differences between plans and outcomes. Indeed, using post implementation empirical data to inform improved sustainability interventions in the future would align with adaptive management paradigms (Folke, Hahn, Ollson, & Norberg, 2005). While it would seem that as social theorists, political ecologists would resist terms like "experiment," their methods–case studies, ethnographies, and participatory action research – seek to explain social and environmental change. These methods of inquiry generate hypotheses about urban dynamics. McKinnon et al. propose that social processes at the urban-rural interface produce at least four patterns of development: land preservation, lawns, small-scale resource production, and urban service-amenity sites. Similar social processes likely contribute to the prevalence of each pattern across exurban sites, an important point of departure for urban sustainability scholarship.

In this features, we have largely focused on issues of urban sustainability and scholarship in the United States. However, the global environmental challenges of rapid population growth and growing inequity constitute a major sustainability challenge of the 21st century (Seto et al., 2010). Cities in the developing world are much more densely populated than those of the West. Such cities house millions living in informal settlements that lack conventional urban infrastructure. Conversely, cities in Europe must confront declining and aging populations, sapping economic vitality but presenting opportunities for re-greening the built environment (Martinez-Fernandez, Audirac, Fol, & Cunningham-Sabot, 2012).

For more than 20 years, geographers have recognized the potential of urban sustainability research but geographic perspective on urban sustainability remain relatively siloed, inhibiting innovative scholarship and obfuscating the contributions to the broader, interdisciplinary communities engaging those themes. To go beyond idiosyncratic research nodes requires broader communication and a good place to start would be to seek out and welcome such scholarship in journals like *Urban Geography*.

Disclosure statement

No potential conflict of interest was reported by the authors.

References

Abler, Ronald A., Adams, John. S., & Gould, Peter. (1971). *Spatial organization: The geographer's view of the world* (No. 910.1 A2).New Jersey: Prentice Hall.
Alberti, Marina. (2017, March 14). Grand Challenges in Urban Science. *Frontiers in Built Environment, 3*. doi:10.3389/fbuil.2017.00006
Al-Hindi, Karen Falconer, & Till, Karen E. (2001). (Re)placing the new urbanism debates: Toward an interdisciplinary research Agenda. *Urban Geography, 22*(3), 189–201.

Arendt, Randall C. (1996). *Conservation Design for Subdivisions: A Practical Guide to Creating Open Space Networks*. Island Press: Washington, D.C.
Ashmore, Peter, & Dodson, Belinda. (2017). Urbanizing physical geography. *The Canadian Geographer, 61*(1), 102–106.
Baerwald, Thomas. J. (2010). Prospects for geography as an interdisciplinary discipline. *Annals of the Association of American Geographers, 100*(3), 493–501.
Berry, Brian JL, & Horton, Frank E. (1974). *Urban environmental management: Planning for pollution control*. Englewood Cliffs, NJ: Prentice Hall.
Berry, Brian. J., & Garrison, William. L. (1958). Alternate explanations of urban rank-size relationships 1. *Annals of the Association of American Geographers, 48*(1), 83–90.
Bettencourt, Luís M. A, & Kaur, Jasleen. (2011). Evolution and structure of sustainability science. *Proceedings of the National Academy of Sciences, 108*(49):19540-19545.
Blaikie, Piers. (2012). Should some political ecology be useful? the inaugural lecture for the cultural and political ecology specialty group, annual meeting of the association of american geographers, april 2010. *Geoforum, 43*(2), 231-239.
Braun, Bruce. (2005). Environmental issues: Writing a more-than-human urban geography. *Progress in Human Geography, 29*(5), 635–650.
Cash, David W., Clark, William C., Alcock, Frank., Dickson, Nancy M., Eckley, N., Guston, David H., ... Mitchell, Ronald B. (2003). Knowledge systems for sustainable development. *Proceedings of the National Academy of Sciences, 100*(14), 8086–8091.
Clark, William C. (2000). Sustainability science: a room of its own. Proceedings of the National Academy of Sciences, *100*(6):1737-1738.
Clark, William. D. (2007). Sustainability science: A room of its own. *Proceedings of the National Academy of Sciences, 104*(6), 1737–1738.
Clark, William. D, & Dickson, Nancy. M. (2003). Sustainability Science focuses on the dynamic interactions between nature and society. *Proceedings of the National Academy of Sciences, 100* (14), 8059–8061.
Cooke, Jason, & Lewis, Robert. (2010). The nature of circulation: The urban political ecology of Chicago's Michigan avenue bridge, 1909–1930. *Urban Geography, 31*(3), 348–368.
Crutzen, Paul. J. (2002). The Anthropocene: Geology of mankind. *Nature, 415*(23), -pp.
Dear, Michael. (2002a). Los Angeles and the Chicago school: Invitation to a debate. *Cities and Community, 1*, 1.
Dear, Michael. (2002b). *From Chicago to LA: Making sense of urban theory*. London: Sage.
Detwyler, Thomas R., & Marcus, Melvin. (1972). *Urbanization and environment: The physical geography of the city*. Duxbury Press: Belmont.
Douglas, Ian. (1981). The city as an ecosystem. *Progress in Physical Geography, 5*(3), 315–367.
Douglas, Ian. (1983). *The urban environment*. Baltimore, MD: Eward Arnold.
Duany, Andres., Plater-Zyberk, Elizabeth, & Speck, Jeff. (2000). *Suburban nation: The rise of sprawl and the decline of the American Dream*. New York: North Point Press.
Ellis, Cliff. (2002). The new urbanism: Critiques and rebuttals. *Journal of Urban Design, 7*(3), 261–291.
Folke, Carl, Hahn, Thomas, Ollson, Per, & Norberg, Jon. (2005). Adaptive governance of social-ecological systems. *Annual Review of Environment and Resources, 30*, 441–473.
Forman, Robert T. (2014). *Urban ecology: Science of cities*. Cambridge: Cambridge University Press.
Frank, John David, & Gabler, Jay. (2006). *Reconstructing the university: Worldwide shifts in academia in the 20th century*. Stanford, CA: Stanford University Press.
Georgescu, Matei, Morefield, Philip E., Beirwagen, Britta G., & Weaver, Christopher P. (2012). Urban adaptation can roll back warming of emerging megapolitan regions. *Proceeding of the National Academy of Sciences, 111*(8), 2909–2914.
Getis, Arthur, & Getis, Judith. (1966). Christaller's central place theory. *Journal of Geography, 65* (5), 220–226.
Grimm, Nancy B., Faeth, Stanley H., Goluiewski, Nancy E., Redman, Charles L., Wu, Jianguo, Bai, X., & Briggs, John M. (2008). Global change and the ecology of cities. *Science, 319*(5864), 756–760.

Hanson, Susan. (1999). Isms and schisms: Healing the rift between the nature-society and space-society traditions in human geography. *Annals of the Association of American Geographers*, 89(1), 133–143.

Hanson, Susan, & Lake, Robert W. (2000). Needed: Geographic research on urban sustainability. *Urban Geography*, 21(1), 1–4.

Harvey, David. (1973). *Social justice and the city*. Baltimore: Johns Hopkins University Press.

Hutchison, Ray. 2010. *Encyclopedia of Urban Studies*. Sage: Thousand Oaks, CA.

Kaplan, David. H., & Holloway, Steven. (2014). *Urban geography*. Hoboken, NJ: Wiley Global Education.

Kates, Robert W., Clark, William C., Corell, Robert, Hall, J. Michael, Jaeger, Carlo C., Lowe, Ian, … Svedin, Uno. (2000). Sustainability science. *Science*, 292(5517), 641–642.

Kates, Robert. (2011). What kind of a science is sustainability science? *Proceedings of the National Academy of Sciences*, 108(49), 19449–19450.

Keil, Roger. (2003). Urban political ecology. *Urban Geography*, 24(8), 723–738.

Keil, Roger. (2013). Progress report—Urban political ecology. *Urban Geography*, 26(7), 640–651.

Ley, David. (1983). *A social geography of the city*. New York: Harper & Row.

Longley, Paul A. (2002). Geographical information systems: Will developments in urban remote sensing and GIS lead to 'better' urban geography? *Progress in Human Geography*, 26(2), 231–239.

Lubchenco, Jane. (1998). Entering the century of the environment: A new social contract for science. *Science*, 279(5350), 491–497.

Magliocca, Nicholas R., Ellis, Erle C., Allington, Ginger R.H., De Bremond, Ariane, Dell'Angelo, Jampel., Mertz, Ole., … Verberg, Peter H. (2018). Closing global knowledge gaps: Producing generalized knowledge from case studies of social-ecological systems. *Global Environmental Change*, 50, 1–14.

Martinez-Fernandez, Christina, Audirac, Ivonne, Fol, Sylvie, & Cunningham-Sabot, Emmanuéle. (2012). Shrinking cities: Urban challenges of globalization. *International Journal of Urban and Regional Research*, 36(2), 213–225.

McHarg, Ian. (1969). *Design with nature*. New Jersey: Wiley.

Meyfroidt, Patrick. (2016). Approaches and Terminology for Causal Analysis In Land Systems Science, *Journal of Land Use Science*, 11(5), 501–522..

Miller, Thaddeus R., Baird, Timothy D., Littlefield, Caitlin M., Kofinas, Gary, Chapin, F. Stuart, III, & Redman, Charles L. (2008). Epistemological pluralism: Reorganizing interdisciplinary research. *Ecology and Society*, 13(2), 46.

Mooney, Harold. A., Duraiappah, Anantha., Lariguaderie, Anne. (2013). Evolution of natural and social science interactions in global change research programs. *Proceedings Of The National Academy Of Science*, 110(Supplement 1), 3665–3572.

Morgan, Grove, J., Cadenasso, Mary L., Pickett, Steward T., Machlis, Gary E., & Burch, William. (2015). *The baltimore school of urban ecology: Space, scale, and time for the study of cities*. Yale University Press: New Haven.

Munroe, Darla K., McSweeney, Kendra, Olson, Jeffrey L., & Mansfield, Becky. (2014). Using economic geography to reinvigorate land-change science. *Geoforum*, 52, 12–21.

Myers, Garth A. (2008). Peri-urban land reform, political-economic reform, and urban political ecology in Zanzibar. *Urban Geography*, 29(3), 264–288.

National Research Council. (1999). *Our common journey: a transition toward sustainability*. National Academy Press, Washington, DC.

Nicholas, Fyfe, & Kenny, Judith T. (2005). *The urban geography reader*. Routledge: London.

Nicholls, Walter J. (2010). The Los Angeles school: Difference, politics, city. *International Journal of Urban and Regional Research*, 35(1), 189–206.

Park, Robert. E., Burgess, Ernest. W., & McKenzie, R. D. (1984). *The city*. Chicago: University of Chicago Press.

Petts, Judith, Owens, Susan, & Bulkeley, Harriet. (2008). Crossing boundaries: Interdisciplinarity in the context of urban environments. *Geoforum*, 39(2), 593–601.

Platt, Rutherford H. (2006). *The humane metropolis: People and nature in the 21st-century city*. Amherst, MA: Univ of Massachusetts Press.

Pred, Allan. R. (1964). The intrametropolitan location of American manufacturing. *Annals of the Association of American Geographers, 54*(2), 165–180.

Quastel, Noah. (2009). Political ecologies of gentrification. *Urban Geography, 30*(7), 694–725.

Quastel, Noah, Moos, Markis, & Lynch, Nicholas. (2012). Sustainability-as-density and the return of the social: The case of Vancouver, British Columbia. *Urban Geography, 33*(7), 1055–1084.

Ranganathan, Malani, & Balazs, Carolina. (2015). Water marginalization at the urban fringe: Environmental justice and urban political ecology across the North-South divide. *Urban Geography, 36*(3), 403–423.

Ramalho, Cristina E. and Hobbs, Richard J. 2012. Time for a change: dynamic urban ecology. *Trends in Ecology & Evolution,* 27(3):179–188

Reid, Ewing, & Cervero, Robert. (2001). Travel and the built environment: A synthesis. *Transportation Research Record, 1780,* 97–114.

Rindfuss, Ronald R., Walsh, Stephen J., Turner, Billie L., II, Fox, Jefferson, & Mishra, Vinod. (2004). Developing a science of land change: Challenges and methodological issues. *Proceedings of the National Academy of Science, 101*(39), 13976–13981.

Robbins, Paul. (2007). *Lawn people: How grasses, weeds, and chemicals make us who we are.* Philadelphia, PA: Temple University Press.

Seto, Karen C., Golden, Jay S., Alberti, M., & Turner, Billie L, II. (2017). Sustainability in an urbanizing planet. *Proceedings of the National Academy of Science, 114 (34)* 8935-8938.

Seto, Karen C, & Reenberg, Anette. (2014). *Rethinking global land use in an urban era.* Cambridge, MA: MIT Press.

Seto, Karen C., Reenberg, Anette., Boone, Christopher G., Fragkais, Michail., Haase, Dagmar., Langanke, Tobias., ... Simon, David. (2012). Urban land teleconnections and sustainability. *Proceedings of the National Academy of Sciences, 109*(20), 7687–7692.

Seto, Karen C., Sánchez-Rodríguez, Roberto., & Fragkais, Michail. (2010). The new geography of contemporary urbanization and the environment. *Annual Review of Environment and Resources, 35,* 167–194.

Shearmur, Richard. (2008). Chicago and L.A.: A clash of epistemologies. *Urban Geography, 29*(2), 167–176.

Skole, David. L. (2004). Geography as a great intellectual melting pot and the preeminent interdisciplinary environmental discipline. *Annals of the American Association of Geographers, 94*(4), 739–743.

Solecki, William, Seto, Karen C, & Marcotullio, Peter J. (2013). It's time for an urbanization science. *Environment: Science and policy for sustainable development.* 55(1), 12–17.

Stokes, Eleanor C, & Seto, Karen C. (2016). Climate change and urban land systems: Bridging the gaps between urbanism and land science. *Journal of Land Use Science, 11*(6), 698–708.

Swyngedouw, Erik, & Heynen, Nikolas C. (2004). *Urban political ecology, justice and the politics of scale.* John Wiley & Sons: London.

Talen, Emily. (2002). The social goals of new urbanism. *Housing Policy Debate, 13*(1), 165–188.

Till, Karen. (2001). New urbanism and nature: Green marketing and the neotraditional community. *Urban Geography, 22*(2), 220–248.

Trudeau, Dan, & Kaplan, Jeffrey. (2016). Is there diversity in the New urbanism? Analyzing the demographic characteristics of New Urbanist neighborhoods in the United States. *Urban Geography, 37*(3), 458–482.

Trudeau, Dan, & Malloy, Paul. (2011). Suburbs in disguise? Examining the geographies of the new urbanism. *Urban Geography, 32*(3), 424–447.

Turner, Billie L, II. (2002). Contested identities: Human-environment geography and disciplinary implication in a restructuring academy. *Annals of the Association of American Geographers, 92* (1), 52–74.

Turner, Billie L, II, Lambin, Eric F., & Reenberg, Anette. (2007). The emergence of land change science for global environmental change. *Proceedings of the National Academy of Sciences, 104* (52), 20666–20671.

Turner, Billie L, II, & Robbins, Paul. (2008). Land-change science and political ecology: Similarities, differences, and implications for sustainability science. *Annual Review of Environment and Resources, 33*, 295–316.

Walker, Paul A. (2005). Political ecology: Where is the ecology? *Progress in Human Geography, 29*, 73–82.

Wehrwein, George S. (1942). The rural-urban fringe. *Economic Geography, 18*(3), 217–228.

Wentz, Elizabeth A., Seto, Karen C., Myint, Soe W., Netzband, Maik, & Fragkias, Michail. (2011). Urban remote sensing (URS) and forecasting urban landuse (FORE) workshops: Common ground and targeted opportunities. *UGEC Viewpoints*, 15–17. Retrieved April 28, 2016, from https://ugec.org/docs/ugec/reports/fore-report.pdf

Wheeler, James O. (1993). Characteristics and recent trends in urban geography. *Urban Geography, 14*(1), 48-56.

Wigginton, Nicholas S., Fahrenkamp-Uppenbrink, J., Wible, B., & Malakoff, D. (2016). Cities are the future. *Science, 352*(6288), 904–905.

Wolch, Jennifer. (2007). Green urban worlds. *Annals of the Association of American Geographers, 97*(2), 373–384.

Yang, Xiaojun (Ed.). (2011). *Urban remote sensing: Monitoring, synthesis and modeling in the urban environment.* Oxford: Wiley-Blackwell.

Zimmerman, Jeffrey. (2001). The "Nature" of urbanism on the new urbanist Fronteir: Sustainable development, or defense of the suburban dream. *Urban Geography, 22*(3), 249–267.

Transportation sustainability in the urban context: a comprehensive review

Selima Sultana ⓘ, Deborah Salon ⓘ and Michael Kuby ⓘ

ABSTRACT
Although the term "sustainability" did not gain traction until the 1980s, concerns about the consequences of transportation technology started long before. This paper reviews the literature on urban transportation sustainability using three frameworks. First, urban transportation can be unsustainable environmentally, economically, and socially (the three pillars of sustainability). Second, sustainable strategies tend to fall into two paradigms. Sustainable Transport Technology improves current patterns of modes and trips by consuming less resources and generating less waste. Sustainable Travel Behavior and Built Environment takes a more holistic approach that targets more sustainable travel choices, recognizing that changes in the built environment that currently constrains those choices are also essential. Third, the Planner's Triangle helps explain commonly encountered situations where inherent tradeoffs can impede win-win-win strategies across environmental, economic, and social domains. The paper concludes with future research directions and concluding thoughts about urban transportation and sustainability.

Introduction

The focus of this review is the sustainability of transportation in the urban context. Transportation is intertwined with many aspects of urban geography and planning, especially when considering sustainability. The Brundtland Commission (1987) defined sustainable development as development that "meets the needs of the present without compromising the ability of future generations to meet their own needs." (p. 8). Although the term "sustainability" was rarely used in urban transportation and urban geography until the early 1990s (Purvis & Grainger, 2004), concerns over the development of transportation systems and their associated impacts on urban form, air quality, and society go back over a century (Alonso, Monzón, & Cascajo, 2015; Carroll & Bovls, 1957; Dickinson, 1949; Kidd, 1992; Marble, 1959; Pratt, 1911; Stradling & Thorsheim, 1999; Taylor, 1915).

Urban transport systems today are largely motorized, dependent on nonrenewable fossil fuels (Black, 2010). Emissions from vehicles contribute significantly to both climate change and the degradation of urban air quality (Chapman, 2007; U.N. Habitat, 2016). Traffic congestion causes substantial economic losses in wasted time and fuel (Downs, 2005). The geographic organization of land use and transport infrastructure in our cities can promote social equity (Curtis, 2008; Van Wee & Handy, 2016) or lead to social exclusion for disadvantaged groups (El-Geneidy et al., 2016; Lucas, Van Wee, & Maat, 2016; Manaugh & El-Geneidy, 2012). New research has begun to show that transport systems affect our happiness (Pfeiffer & Cloutier, 2016) and overall quality of life (Bäckström, Sandow, & Westerlund, 2016; Bergstad et al., 2011; Friman, Fujii, Ettema, Gärling, & Olsson, 2013).

This paper reviews the literature on urban transportation sustainability using three frameworks. First, the goals of transportation sustainability are summarized in three fundamental "pillars" of environmental, social, and economic quality (Gudmundsson, Hall, Marsden, & Zietsman, 2016). The unsustainability of urban transportation systems can also be understood via these three pillars. Second, the literature on sustainable transport solutions is often divided into narrow and broad approaches (Litman & Burwell, 2006), as summarized in Figure 1. The narrower Sustainable Transportation Technology approach focuses on making each form of mobility more sustainable by reducing its resource use and pollution. The broader Sustainable Travel Behavior and Land Use approach is more holistic. It recognizes that moving people and goods more sustainably will require a reconfiguration of urban form to improve accessibility for more sustainable transport modes. While both approaches address concerns about the unsustainability of our current transportation system, they emphasize different solutions.

The third framework is based on Campbell's Planner's Triangle (1996, 2016), which emphasizes the sometimes inherent conflicts and tradeoffs between economic, social, and environmental goals (the three pillars of sustainability). Win-win-win solutions are

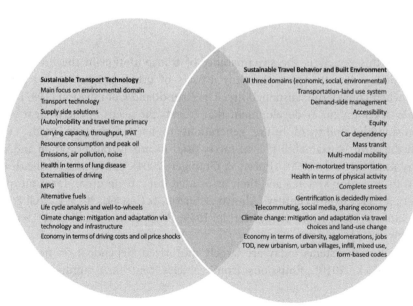

Figure 1. Two main approaches to urban transportation sustainability.

elusive; efforts to promote one or two pillars of sustainability often conflict with the third. We offer hopefully thought-provoking examples of typical and often inevitable conflicts between sustainability goals. Our contribution is an up-to-date, comprehensive review of the literature on urban transportation sustainability. We use three distinct frameworks to organize this literature, provide numerous examples of the challenges of urban transportation sustainability, and conclude with a long list of promising future research directions.

Unsustainability of urban transportation

The study of urban transport sustainability starts by identifying what makes urban transport *un*sustainable. We present a brief overview of the main problems, each of which impacts the three pillars of sustainability differently (see Table 1 for a summary). Due to space limitations, concerns such as noise, wetlands, water pollution, wildlife, agriculture, land-surface modification, and archaeological resources are not covered.

Material throughput and carrying capacity

The term "sustainability" has been credited to Meadows, Meadows, Randers, and Behrens (1972) global simulation classic, *The Limits to Growth*. Accordingly, sustainability in transportation was initially viewed from the principles of environmental carrying capacity (Black, 1997a). Three conditions were identified for meeting the transport needs of future generations (Black, 1997b): 1) renewable resource use not exceed their rates of regeneration; 2) non-renewable resource use not exceed the rate at which sustainable renewable substitutes are developed; and, 3) pollution emissions not exceed the assimilative capacity of the environment. The IPAT identity was an early organizing principle for understanding the driving forces of population (P), affluence (A), and technology (T) that lie behind the activities that generate the material throughput (I) that causes environmental impacts (Chertow, 2000). In the transportation context, an example might be:

$$I(\text{barrels of oil consumption}) = P(\text{persons}) \times A(\$GDP/\text{person}) \times T(\text{barrels}/\$GDP)$$

In a similar vein, the transport-specific ASIF framework (Schipper, 2011) estimates transport sector carbon emissions as a function of A (volume of transport activity), S (mode share), I (energy per ton-mile or passenger-mile), and F (carbon content of the fuel). Since material throughput includes both resource inputs and waste outputs, its impacts listed in Table 1 include the primary effects of the next three categories: oil supply, air pollution, and climate change.

Oil supply, reserves, and prices

The oil crises of the 1970s were seminal events in transportation sustainability. Precipitated by geopolitical events in the Middle East and the formation of OPEC, inflation-adjusted crude oil prices increased 10-fold (British Petroleum, 2017). From the

Table 1. Sustainability impacts of urban transportation.

Unsustainable Transportation-Related Problem	Economic	Social	Environmental
Material throughput and carrying capacity	Primary – see oil supply, reserves, and prices	Primary – see air pollution and climate change	Primary – see air pollution and climate change
Oil supply, reserves, and prices	Primary – oil prices	Secondary – price effects on families and communities	Secondary – drilling in sensitive ecoregions
Air pollution	Secondary – health care and other costs	Primary – health	Primary – air quality
Climate change	Primary – impacts on all economic sectors	Primary – e.g., health, occupations, food, safety, water, migration, conflict	Primary – e.g., biodiversity, precipitation, extreme weather
Traffic congestion	Primary – time costs	Secondary – time constraints on households	Secondary – increased emissions
Road safety	Primary – accident costs	Primary – injuries and deaths	Secondary – resources used in repairing and replacing
Transportation affordability	Primary – accessibility to jobs, school, etc.	Primary – household budgets	Secondary – higher emissions of older cars
Equity	Secondary – exclusion of potential consumers and workers	Primary – accessibility to jobs, school, etc.	Secondary – equal access to alternatives to driving will improve environmental quality
Physical activity and health	Secondary – medical costs	Primary – health	Secondary – zero emissions of non-motorized modes

1970s to the 2008 financial crisis, the problems associated with finite oil reserves returned in repeated cycles: the volatility of oil prices, the vulnerability of the economy to recession and inflation, the inelasticity of oil demand, the boom and bust of oil exploration and production, the military costs of keeping oil shipping lanes open, and the pressure to drill in sensitive offshore and wilderness locations. In the 2000s, these concerns were joined by the rapid motorization of newly industrializing countries, especially China (Sperling & Gordon, 2009). By the mid-2010s, however, the fears of permanent $150 per barrel crude prices and $5 per gallon gasoline prices in the United States (Mann, 2013; White, 2011) had receded into the future due in part to the invention of hydraulic fracturing ("fracking") technology.

Air pollution

In the U.S., transportation is responsible for over 20% of ultrafine particulate matter, over 30% of volatile organic compounds, and over 50% of nitrogen oxide emissions (USEPA, 2016). Concentrations of these pollutants are substantially elevated near major roads (Karner, Eisinger, & Niemeier, 2010). In the U.S. alone, air pollution from transportation causes hundreds of thousands of early deaths (Caiazzo, Ashok, Waitz, Yim, & Barrett, 2013). U.S. air quality regulations that reduce this exposure are estimated to account for 61–80% of the monetized benefits of all federal regulations (USOMB, 2016).

Climate change

Transportation both impacts, and is impacted by, climate change. In March 2016, transportation eclipsed power generation as the leading source of CO_2 emissions in the United States, and is the only energy-use sector with increasing emissions (USDOE, 2016). The effects of climate change – extreme storms and associated flooding, sea-level rise combined with storm surges, higher temperatures, melting permafrost, and wildfires – will directly affect transportation infrastructure and vehicles and indirectly affect traveler behavior, population shifts, and supply and demand for freight (National Research Council, 2008). The Third National Climate Assessment adopted a risk-management approach to assess the risks in terms of (a) likelihood of occurrence and (b) magnitude of impact (McLaughlin, Murrell, & DesRoches, 2011; Schwartz et al., 2014). For instance, inundation of coastal roads, rail, and airports is considered (a) virtually certain to occur and (b) highly disruptive.

Traffic congestion

Delay from traffic congestion is a large external cost to our urban transport system. The total cost of this delay includes the cost of fuel, added pollution, and time wasted by commuters and freight stuck in traffic. A recent report by the Centre for Economics and Business Research (2014) estimated that the total cost of traffic congestion in the United States reached $124 billion in 2013, and projected that traffic congestion costs would increase an additional 50% by 2030. By many accounts, congestion is the single largest external cost of our transport system (e.g. Parry, Walls, & Harrington, 2007).

Road safety

Annually, 1.25 million people die in crashes on the world's roads (World Health Organization, 2016). Traffic-related deaths are the leading cause of mortality for 15–29 year-olds worldwide (World Health Organization, 2016). In U.S. cities, bicyclists are 12 times more likely to be killed than drivers (Brustman, 1999; Buehler & Pucher, 2017). Even though bicycle and pedestrian deaths have declined since 1975, each year bicyclists account for 2% of deaths and pedestrians for 15% (IIHSLDI, 2016). The U.S. has much higher fatality and serious injury rates per kilometer driven than other developed countries (Buehler & Pucher, 2017). Urban fatality rates per million VMT are 62% lower than rural rates (NHTSA, 2017). Many cities worldwide have adopted Vision Zero with the goal of eliminating traffic deaths altogether (Vision Zero, 2017).

Transportation affordability

Transport affordability is a substantial challenge for many urban residents in both the developed and developing worlds. In the U.S., transportation is the second highest expenditure category after housing (BLS, 2015). Low-income households choose between struggling to pay for vehicle ownership (e.g., Puentes & Roberto, 2008) and struggling to access jobs and services without a car (e.g., Kawabata & Shen, 2007). In much of the developing world, urban residents choose between struggling to pay transit fares and traveling on foot (e.g., Salon & Gulyani, 2010).

There is an important spatial tradeoff at work here for urban residents. As predicted by economic theory (Alonso, 1964), housing costs generally increase with accessibility. All else equal, housing and transport costs move in opposing directions across space in a metropolitan area. Lower-income households may choose to live on the urban fringe in order to be able to afford housing – especially if they are interested in home ownership – and then be stuck with high transportation costs (Haas, Newmark, & Morrison, 2016; Kane, York, Tuccillo, Gentile, & Ouyang, 2014; Mattingly & Morrissey, 2014).

Equity

Transport equity is the fairness with which benefits and costs of transportation are distributed (Litman, 2016). Transport inequity results when costs paid are not proportional to benefits received (Taylor, 2017). Looking at benefits only, transport inequity occurs when a city's transportation-land use system affords different levels of access to different groups. Transport inequity can lead to social exclusion (Lucas, 2012), which results in the inability of certain groups to fully participate in the economic and social life of their city.

Equitable transportation systems need not provide equal services to all areas and socioeconomic groups. Indeed, the greatest benefits for certain transport investments can go to the least-advantaged areas and groups based on need. For instance, greater investment in public transit, pedestrian, and bicycle facilities in lower-income communities is warranted because residents of these areas own fewer cars.

While much has been written about social equity and urban transport (Curtis & Low, 2016; Preston, 2009), integrating equity into transportation planning practice remains rare (Manaugh, Badami, & El-Geneidy, 2015). Since transportation investments historically created spatio-temporal disparities in access to opportunities (Lucas, 2012), remedying this should be a high priority.

Physical activity and health

Active transportation promotes health (Bertolini, Le Clercq, & Kapoen, 2005; Sallis et al., 2016). Physical inactivity is the fourth leading risk factor of chronic diseases such as cancer, heart disease, and type-2 diabetes (WHO, 2010). Inactivity accounts for 3.2 million deaths each year (Pratt, Norris, Lobelo, Roux, & Wang, 2014) and over $50 billion in health-care costs worldwide (Ding et al., 2016). In the U.S., 8.7% of health care costs are for treating diseases associated with physical inactivity (Carlson, Fulton, Pratt, Yang, & Adams, 2015). By almost any measure, however, active transportation represents a small fraction of all travel – at least in the U.S. Less than 4% of Americans walk or bike to work (Sultana, 2014). The National Household Travel Survey (2009) reports less than 35% of Americans walked at all in the previous week.

Sustainable transportation technology

The most direct approach to improving transport sustainability, but also the most narrowly defined, is to make the same trips on the same modes more sustainable (Black, 1997b). Techno-centric approaches recognize that American cities and lifestyles have evolved to the point that they revolve around the automobile and view this dependency as extremely difficult to change in the short or medium term (Sperling & Gordon, 2009). Sustainable Transportation Technology solutions include policies that address innovation, infrastructure investment, energy efficiency, alternative fuels, pollution control, and intelligent transportation systems (Huang, Kuby, & Chow, 2017). Additional emerging technologies are discussed later as promising research directions.

Cleaner petroleum-powered cars

To the driving public, transportation sustainability is mainly about fuel-efficient cars. In the U.S., the Corporate Average Fuel Economy (CAFE) standards are the primary policy to increase fuel economy (Greene, 1998). In other countries, high fuel taxes play a much larger role in encouraging efficient cars (and fewer and shorter car trips). Regulatory mechanisms requiring pollution-control equipment have dramatically reduced emissions of harmful pollutants from vehicles – although these technologies do not reduce CO_2 emissions (Godish, Davis, & Fu, 2014). A major advantage of focusing on cleaner and more efficient petroleum-powered cars is that it requires no revolutionary changes in travel behavior, urban form, technology, or infrastructure.

Alternative fuels, stations, and vehicles

The U.S. relies on oil for over 92% of its transportation energy. To reduce oil dependence and meet the Paris Agreement's goal of 80% reduction of greenhouse gas (GHG) emissions from 2005 levels by 2050, greener alternatives to petroleum are needed for motor vehicles. Petroleum, however, enjoys major advantages over alternatives, including energy density, economies of scale, familiarity, and nearly ubiquitous refueling stations. For biofuels, natural gas, electricity, and hydrogen to compete, multiple industries must develop simultaneously, including vehicle and fuel production, fuel retailing, repair, and insurance (Farrell, Keith, & Corbett, 2003; Lu, Rong, You, & Shi, 2014; Melendez, 2006).

A sophisticated literature simulates the roll-out, competition, coordination, economies of scale, learning, and incentives required to transition to greener sources of transportation energy (Fan et al., 2017; Greene, Leiby, & Bowman, 2007; Ogden, Fulton, & Sperling, 2016; Struben & Sterman, 2008). How many stations and where to locate them is an inherently geographical question (Agnolucci & McDowall, 2013; Kuby & Lim, 2005; Melaina & Bremson, 2008), as is the revealed behavior of drivers choosing where to refuel in the face of a scarcity of stations (Kelley & Kuby, 2013; Sperling & Kitamura, 1986). Driving range is a critical constraint to adoption of electric vehicles in particular (Pearre, Kempton, Guensler, & Elango, 2011). Finally, all alternatives to petroleum are not equally economical or sustainable: well-to-wheels and life-cycle analysis evaluate the potential of different energy source-carrier-propulsion pathways (Edwards, Mahieu, Griesemann, Larivé, & Rickeard, 2004; Von Blottnitz & Curran, 2007).

Intelligent transportation systems

Intelligent transportation systems (ITS) include a variety of technologies to manage congestion, including lane sensors, ramp metering, dynamic messaging, and emergency dispatching (Zhang et al., 2011). Recurrent bottlenecks driven by a systemic imbalance of vehicles and road space account for 40%, crashes and disabled vehicles for 25%, weather for 15%, construction for 10%, poor signal timing for 5%, and special events for 5% of congestion (Cambridge Systematics, 2005). Different ITS technologies address each of these. While ITS is generally the province of traffic engineers, it is important for transport and urban geographers to understand because it represents an attractive option for moving more vehicles faster without new road construction or lane expansion.

Sustainable travel behavior and the built environment

Policies, transportation investments, and the built environment affect the demand side of travel by changing the underlying price, time, comfort, and convenience of transport alternatives. Built environment and transport investment strategies can affect transport sustainability through (1) reducing trip distances, (2) reducing trip frequencies, (3) increasing vehicle occupancy, or (4) shifting travel to non-car modes. Table 2 presents a matrix illustrating how a variety of policy and investment strategies lead to changes in the transportation and built environment system that affect travel behavior in each of these four ways. Plus and minus signs indicate the expected directions of the effects. Notice that some changes may have conflicting effects on sustainability. This list is not comprehensive: we do not cover carpooling, telecommuting, online shopping and education, social media, and behavioral-change programs.

Table 2. How transport/land-use system changes impact travel demand.

System Change	Policy/Investment Actions (examples)	Trip Distance	Trip Frequency	Carpooling (vehicle occupancy)	Non-Auto Mode
Increase density	• Multifamily housing development • Increase housing unit density • Urban infill development	−	+	+	+
Mix land uses	• Mixed-use zoning • Vertically mix buildings	−	±	+	+
Increase local access to jobs	• Jobs-housing balance • Mixed-use zoning • Higher floor-area ratio • Reduce parking requirements	−	+		+
Increase regional access to jobs	• Economic development strategies such as small business incentives and support services, and business-enabling zoning changes	+/−			
Improve network connectivity	• Reduce block length • Grid network design • Cut-through streets and paths	−	+		+
Improve public transport access	• Add transit routes • Increase service frequency • Transit-oriented development • Intermodal infrastructure (e.g. bikeshare, bike racks)				+
Improve public transport service	• Real-time arrival information to stations and stops • Premium (e.g., faster, more comfortable) service for an additional charge • Additional amenities on transit vehicles and at transfer hubs (e.g. wi-fi)				+
Improve walkability	• Complete streets • Sidewalks and paths • Road diets • Traffic calming • Safer crossings				+
Improve bikeability	• Complete streets • Bicycle lanes, paths, and boxes • Bicycle boulevards • Road diets • Traffic calming • Bicycle parking				+
Reduce parking supply	• Modify off-street parking minimums and maximums • Convert on-street parking to BRT, streetcar, or bike lanes		−	+	+

(Continued)

Table 2. (Continued).

System Change	Policy/Investment Actions (examples)	Trip Distance	Trip Frequency	Carpooling (vehicle occupancy)	Non-Auto Mode
Increase prices	• Raise fuel taxes • Raise parking fees • Raise licensing and registration fees • Road user fees • Congestion pricing • Carbon tax	+	+	-	-
Increase highway capacity	• Build new roads or widen existing ones • Traffic management	+	+	-	-

Pricing

One of the most powerful tools to encourage behavior change is pricing. When goods are more expensive, people voluntarily consume less of them, and transportation is no exception. The literature is clear on this. The price of fuel affects both the fuel efficiency of the vehicles purchased (Busse, Knittel, & Zettelmeyer, 2013) and how much people drive (Gillingham, 2014; Goodwin, Dargay, & Hanly, 2004; Graham & Glaister, 2004; Lane, 2010). Road pricing – or its more sophisticated cousin, congestion pricing – is identified as possibly the only policy tool that can substantially reduce traffic congestion (Downs, 1992, 2004, 2005). Transit fares impact ridership (Taylor, Miller, Iseki, & Fink, 2009). Raising parking prices reduces parking occupancy (Ottosson, Chen, Wang, & Lin, 2013; Pierce & Shoup, 2013).

Modes

An essential part of making urban systems more sustainable is providing viable alternatives to cars. Historically, the primary alternatives included transit, walking, and bicycling. Transit provision both reduces car dependence (Spears, Boarnet, & Houston, 2016) and increases transit use (Taylor et al., 2009). Pedestrian-friendly neighborhoods encourage walking (Saelens & Handy, 2008) and bicycle facilities strongly encourage biking (Dill & Carr, 2003; Pucher, Dill, & Handy, 2010). Evidence on the extent to which active travel reduces car use, however, is weaker (Piatkowski, Krizek, & Handy, 2015).

The "alternatives" to cars have recently expanded to include ways of accessing cars without owning them, such as car-sharing organizations (e.g., Zipcar) and ride-sourcing smartphone apps (e.g., Uber, Lyft). While trips made using these services are car trips, the sustainability advantage is that they allow households to own fewer cars and make more of their other trips using alternative modes. There is convincing evidence that car-sharing organizations measurably reduce car ownership (Martin, Shaheen, & Lidicker, 2010). The evidence on ride-sourcing is less well-developed, but early studies suggest ride-sourcing may improve urban transport sustainability (e.g., Li, Hong, & Zhang, 2016; Rayle, Dai, Chan, Cervero, & Shaheen, 2016).

Land use

Land-use strategies ranging from leapfrog, beltway, and exurban development at one end of the spectrum to infill, urban villages, smart growth, and new urbanism at the other can affect transport system sustainability in either direction. They do so by bringing trip origins and destinations closer together (e.g., jobs/housing balance, density), forcing them farther apart (e.g., large lot zoning), diversifying neighborhoods (e.g., mixed-use development), improving transit access (e.g., transit-oriented development), and setting minimum or maximum off-street parking requirements.

There are large differences in car dependence across built environment typologies, regardless of scale. At the metropolitan scale, average car use in 13 major U.S. cities was 7.5 times greater than in Singapore, Tokyo, and Hong Kong (Kenworthy & Laube, 1999). Using a neighborhood typology approach, Salon (2015) demonstrated that car dependence differs by a factor of two across neighborhood types in California. Across neighborhoods in San Francisco, car mode share ranged from 60% to 90% (Kitamura, Mokhtarian, & Laidet, 1997).

This evidence, however, does not conclusively show that the built environment has a strong effect on travel behavior. There is the distinct possibility that households that prefer not to drive self-select into neighborhoods and cities that provide good alternatives to driving, and vice versa. A substantial literature (including Kitamura et al., 1997; Salon, 2015) attempts to control for residential self-selection and focuses attention on the direct effect of built environment characteristics on travel behavior.

Reviews of this literature suggest that the built environment does impact travel choices, but that socioeconomic factors are probably more important than built environment characteristics (e.g., Ewing & Cervero, 2010; Salon, Boarnet, Handy, Spears, & Tal, 2012; Sultana, 2015; Sultana & Weber, 2007). Built environment characteristics such as population density, accessibility of employment and other destinations, and land-use diversity are important determinants of travel choices in at least some published studies, and the relationships are in the expected direction (Table 2). A separate literature investigates the effect of parking supply on car dependence, finding strong evidence that more parking spaces lead to more driving (Chester, Fraser, Matute, Flower, & Pendyala, 2015; Guo & Ren, 2013; Sultana, 2015).

Networks

Connectivity and design are the main network features that influence urban sustainability. Connectivity is a distinct concept from accessibility, which factors in the destinations within reach. Connectivity refers to the ability of the network configuration to provide direct routes and options for alternative routes (Berrigan, Pickle, & Dill, 2010). Route-directness (Stangl, 2012) measures the detour required for a particular O-D pair, while aggregate measures such as link-node ratio, intersection density, and block length estimate the general ability to connect origins and destinations (Tal & Handy, 2012). Grid networks generally connect better than post-war suburban networks (Randall & Baetz, 2001), and are strongly associated with more walking and biking (Ewing & Cervero, 2010). Design refers to qualitative aspects of networks, such as wide sidewalks, painted bike lanes, plazas, shade, landscaping, lighting, cleanliness, and amenities, which make each unit of distance travelled safer and more pleasant (Cervero & Kockelman, 1997). The Complete Streets movement emphasizes sustainable design, with over 1,000 local and state agency adoptees (Smart Growth America, 2017).

Conflicting goals: win-win-win solutions are elusive

> "In a world in which the distribution of mobility is enormously unequal, it is inevitable that competing imperatives of reducing greenhouse gas emissions and improving mobility will come into conflict." (Wachs, 2010, p. 9)

It is often difficult to find solutions that simultaneously advance all tenets of sustainability: environmental, economic, and social. Urban planners have long

Figure 2. Planners triangle, adapted from Campbell (1996).

recognized this tension, depicting it graphically as "The Planner's Triangle" (Figure 2) (Campbell, 1996). The triangle highlights the often necessary choice between investing scarce resources toward one sustainability goal at the expense of another. Campbell's 1996 paper was revisited in a series of articles in the *Journal of the American Planning Association* on its 20th anniversary (e.g., Campbell, 2016; Hirt, 2016).

Here, we illustrate some of the challenges outlined by Campbell and others using commonly encountered situations faced by urban decision makers at the intersection of urban sustainability and transport sustainability. These are not case studies per se, but rather are illustrative scenarios intended to inspire creative thinking about controversial urban transport sustainability solutions.

Transit-oriented development and gentrification

Transit-oriented development (TOD) is heralded by many as an important part of the solution to our urban sustainability challenge, improving accessibility and reducing car dependence. TOD is defined as "compact, walkable, pedestrian-oriented, mixed-use communities centered around high quality train systems" (Transit Oriented Development Institute, 2017), and is being built in central urban neighborhoods all over the U.S.

Accessible urban forms, however, attract higher income residents and increase land and housing costs, which can lead to frustration and relocation of low-income renters (Koschinsky & Talen, 2016). Despite the goal of income diversity in these neighborhoods, TOD often worsens social exclusion by displacing the low-income population to neighborhoods that have less walkable access to activities (Pollack, Bluestone, & Billingham, 2010). If affordability is prioritized, there is a potential to reduce social inequality in highly accessible neighborhoods. Unfortunately, prioritizing affordability reduces developer profit and is therefore not simple. Using the terminology of the Planner's Triangle, TOD investments are consistent with both economic and environmental goals, but sometimes will reduce social equity, thereby creating both property and development conflicts.

Infill development and affordable housing provision

Infill development (including most TOD) is both riskier and more expensive than greenfield development on the urban fringe. First, multifamily housing development is inherently riskier than single-family tract development. In single-family developments, homes are sold individually as completed; if demand drops, construction can pause. Multifamily housing developers, however, must invest in entire buildings before any units can be sold or leased; if demand drops, the developer risks substantial losses. Further, developing in central areas requires difficult parcel assembly (Brooks & Lutz, 2016; Terrence, 2001), the possibility of brownfield cleanup requirements (Terrence, 2001), and facing neighborhood opposition (Cervero et al., 2004; McConnell & Wiley, 2010; Terrence, 2001). For these reasons, financing infill development can be challenging and expensive (Venner & Ecola, 2007). In addition, cities often want affordable housing included in infill projects to fulfill social equity goals. However, requiring affordable housing reduces the profit margins on already risky projects, making this type of (more) sustainable development less financially feasible. In Planner's Triangle terms, there is a tradeoff between social equity and the profitability of infill development, creating a property conflict. In addition, infill development is economically riskier than greenfield development, creating a potential resource conflict with infill's clear environmental benefits.

Transit's multiple purposes

Transit provides a more sustainable alternative to the private car, allowing large numbers of people to move through the city without the substantial internal and external costs of driving. Transit – especially rail – can play an important role in channeling economic growth to station neighborhoods (e.g., Mohammad, Graham, Melo, & Anderson, 2013). Transit also provides vital mobility and access services for the urban poor and carless (Giuliano, 2005). Thus, it appears that perhaps public transport investment is a win-win-win solution, improving environmental performance, local economies, and social equity.

The challenge arises when geography is introduced. Transit serves low-income people, aims to get wealthier people out of their cars, and aims to spur local economic development. Giuliano (2012) identified that not only are the poor located in different neighborhoods from the wealthy, but they also have different transportation needs: "Every investment represents a choice of what transit market is to be served" (576). Indeed, some of the most well-known environmental justice lawsuits have centered on the choice of technology (bus vs. rail) and corridor (e.g., Golub, Marcantonio, & Sanchez, 2013; Grengs, 2002). In terms of the Planner's Triangle, this scenario differs from the other examples, in that transit investments generally advance all three goals, but depending on which technology is built in which corridor, one of the goals may lag behind the others, creating property, development, or resource tensions.

Alt-fuel vehicle costs and subsidies

Alternative-fuel vehicles (AFV) are more expensive than their conventional equivalents and are marketed to wealthier, more highly educated, multi-car households, creating tension with equity goals. The success of the AFV industry depends on early adopters to purchase new vehicles and help the industry get up to scale, which will make the vehicles and fuel more affordable to later adopters. For this reason, many governments subsidize early adopters through tax credits, reduced registration fees, and HOV lane access (Alternative Fuels Data Center, 2017), and cluster stations in wealthier target neighborhoods (Ogden & Nicholas, 2011). Some politicians have critiqued EV incentives as poorer citizens subsidizing the wealthy – an argument that is also used against rooftop solar, bicycle infrastructure, and high-speed rail. Recently, some states have not only eliminated subsidies for EV buyers but also raised EV registration fees to compensate for the fact that they do not pay their fair share of road maintenance costs through fuel taxes (Spector & Pyper, 2017). With positive effects on the environment and the economy and negative effects on social equity, subsidizing AFVs creates both property and development conflicts.

Promising research directions

In light of the recent discourse about transport sustainability within the urban context, we now turn to considering what usefully comes next. This penultimate section suggests promising directions for research. These suggestions are not entirely novel; the contribution here is to put them in one place.

Natural and policy experiments to study land use effects on travel behavior

Despite a sizable literature, evidence on the relationship between land-use policy and travel behavior remains weak. Much of the uncertainty stems from the fact that most studies are based on cross-sectional data, collected from many people, but only at one point in time. Although we can and do try to use these data to shed light on the likely effect of changes in Table 2 strategies on car dependence and urban sustainability, there simply is no way for even the most sophisticated statistical methods to fully overcome both the challenges of correlation between variables and to control for neighborhood self-selection. To measure the effects of a change in policy, we need to collect data on travel choices before and after real changes happen, and compare them to estimate the effect. This suggestion is not new (e.g., Boarnet, 2011; Salon et al., 2012), but research employing this evaluation model remains uncommon in the transportation sustainability literature. Recent examples include Spears et al. (2016), Brown et al. (2016), Ye, Mokhtarian, and Circella (2012), and Lovejoy, Sciara, Salon, Handy, and Mokhtarian (2013). Not every land-use change can be studied this way – some effects occur over long periods of time, making this approach infeasible. However, in cases that can be evaluated over a short time horizon, before-after data collection and analysis is a promising avenue for research.

Effects of urban and transportation policy packages

Policymakers often consider policy and investment *packages* rather than actions in isolation. The importance of integrated transport sustainability strategies is acknowledged by the research community (May, Kelly, & Shepherd, 2006), but few studies have quantified the interaction effects. Rare examples include Lee and Lee (2013) and Guo, Agrawal, and Dill (2011). Most quantitative studies have focused almost exclusively on the effects of each strategy individually.

Understanding heterogeneity

Much of the empirical work on travel behavior presents results as though the relationships of interest – for instance the price elasticity of gasoline or the effect of improved employment access on travel – are constant across people and space. These relationships are not constant, however, and understanding the heterogeneity in these relationships will better inform decision makers about how their actions may yield different results for different categories of people, places, and trips. A few recent studies have begun to explore this heterogeneity (e.g., Boarnet, Houston, Ferguson, & Spears, 2011; Salon, 2015), but more work is needed.

Include preferences, experiences, and attitudes in behavioral analysis

How do preferences, past experiences, and attitudes affect travel and related choices? Evidence to date suggests that these factors can add substantial explanatory power to travel behavior models (Sultana, 2015). Parkany, Gallagher, and Viveiros (2004) provide an early review of this literature; Van Acker, Van Wee, and Witlox (2010) provide a conceptual framework for this research area. However, because most surveys do not include questions about preferences, experiences, and attitudes, most analyses do not include them.

Urban freight

Understanding the sustainability implications of urban freight movement in the e-shopping era is an emerging and important area of study. Brick-and-mortar stores are shutting their doors as retail moves online. Research has shown that the environmental impacts of this change in last-mile goods movement are surprisingly ambiguous, and depend critically on the spatial configuration of both the urban environment and the goods distribution network (Wygonik & Goodchild, 2016). At the same time, urban streets and neighborhoods are not built to accommodate large trucks, and "complete streets" designers often neglect to include freight movement in sustainable urban plans (Smart Growth America, 2017; Zavestoski & Agyeman, 2014). Enterprising companies are experimenting with a variety of delivery modes (e.g., bicycles, robots, and drones) and options (e.g., differential pricing by delivery speed, delivery to lockers instead of individual homes) – all of which are ripe areas for research.

Cultural and development-level differences

Research in different cultural and developmental contexts offers opportunities to learn how sustainability policies translate between more- and less-developed countries. There are at least three frames for this topic, all understudied. First, movement toward sustainability means different changes in different contexts. For instance, most megacity dwellers in developing countries would likely view densification as less sustainable and desirable, while the opposite would be true for most U.S. cities (McFarlane, 2016; Moroni, 2016). Second, little is known about the diffusion of transportation technologies from less- to more-developed cities. While cars, trucks, planes, and trains originated in the industrial north, the Global South has led the way with flexible, lower-investment options, such as electric bicycles, bus rapid transit, and informal jitney services. Third, many have urged the developing world to learn from the unsustainable mistakes of the north and "leapfrog" ahead with locally appropriate sustainable land-use policies and infrastructure investments. It is crucial to investigate these successes and failures.

Disadvantaged neighborhoods

Much equity research has compared transportation behavior and service between lower and higher income neighborhoods, but there is a need for in-depth studies of how to better serve the most disadvantaged neighborhoods. It is important to identify particular at-risk groups in close collaboration with the intended population, local policymakers and key stakeholders (Lucas et al., 2016). Some research has explored independent transportation for disadvantaged children (Veitch et al., 2017) and accessibility in disadvantaged neighborhoods (El-Geneidy et al., 2016; Pyrialakou, Gkritza, & Fricker, 2016), but more work is needed.

Equity

Measuring social equity will also require creating novel methodologies by adopting quantitative, qualitative, and participatory approaches (Pereira, Schwanen, & Banister, 2017). Analyses of the distributive effects of transit service changes, for instance, usually relies on the demographics of the service area, with the assumption that the demographics of the ridership is similar, which may not be the case (Karner, Kuby, & Golub, 2017).

High speed rail and hyperloop

Long-distance, high-speed, low-carbon transportation modes have excellent potential for win-win-win outcomes. High-speed rail (HSR) with typical speeds of 250–350 km/h are much faster than automobile and conventional rail and are competitive with airlines over distances from 400–2000 km (Kim, 2016; Li & Loo, 2016). Though still in its conceptual and developmental stage, Elon Musk's idea for a long-distance Hyperloop using vacuum tube technology and small pods is being taken seriously in many circles, including by USDOT (Taylor, Hyde, & Barr, 2016). Both HSR and Hyperloop stations can be located in central business districts, where they will enjoy a large door-to-door travel time advantage over air travel and reinforce the economic vibrancy of central cities (Kim & Sultana, 2017; Vickerman, 2015). Analysis of corridor suitability, direct vs. indirect connectivity, and comparison of door-to-door travel times among long-distance modes for different regions of the city are promising areas for geographical research.

Psychological well-being and happiness

Much attention has been given to the relationship between the built environment and physical health, but more research is needed on its effect on psychological well-being (Galea, Ahern, Rudenstine, Wallace, & Vlahov, 2005). A few studies have focused on built and social environments and their relationship with older adult mental health (Galea et al., 2005) and young adults' quality of life (Xiong & Zhang, 2016). This is an area of research where heterogeneity across socio-economic groups, gender, race, ethnicity, age and place may be especially important.

Alternative-fuel vehicle adoption and station arrangement

Consumer decision-making about AFV purchasing is most often studied using stated preference surveys and experiments (e.g., Bunch, Bradley, Golob, Kitamura, & Occhiuzzo, 1993; Potoglou & Kanaroglou, 2007). Because fuel station availability is but one factor of many, surveys usually present it to respondents in a simplified way in terms of a single station either close to home or along a commuting route, or as a percentage of existing stations offering the fuel. Research is needed on how potential early adopters evaluate the geographic arrangement of *multiple* stations to provide an improved behavioral basis for station planning.

Non-electric AFVs

Most AFV research to date has focused on the arrival of EVs and the potential of fuel-cell vehicles. More research is needed on the purchase and fueling of propane, natural gas, and biofuel vehicles. Millions of Americans own flex-fuel vehicles they can fill with either gasoline or E85 (85% ethanol, 15% gasoline), but little is known about who actually fills them with E85, and when, where, and why. In addition, most AFV research focuses on consumers, but for these fossil and bio alt-fuels, commercial and government fleets are often the early adopters, and fleet vehicles drive more miles per year and use more fuel per mile (Nesbitt & Sperling, 1998).

Connected and autonomous vehicles (CAVs)

Research directions abound as to how CAVs will affect urban sustainability. The primary driving force behind the push for CAVs is increased safety (Anderson et al., 2016), but questions remain about how CAVs will interact with human drivers and whether improved safety will encourage more bicycle riders. In terms of land use, what will be the effect on residential choices and sprawl? Which ownership model (shared vs. private) will dominate, and if shared, which kinds of companies will emerge as the

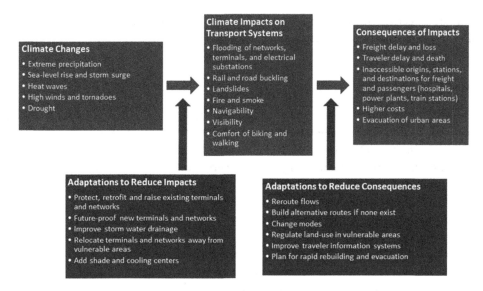

Figure 3. Adaptive strategies to reduce climate change impacts on transportation systems and resulting consequences. Based on Third National Climate Assessment, Transportation chapter (Schwartz et al., 2014).

leaders? The contenders include legacy taxi companies; ride-hailing companies like Uber and Lyft; IT companies like Google or Apple; car manufacturers like GM or Tesla; or possibly even electric utilities, which could use autonomous EVs to store and release excess renewable energy to improve grid stability and operate the transport system under a public-utility model. If one of the shared ownership models wins out, will developers planning new and supposedly more sustainable multi-family housing today find that they were required by outdated code to overbuild their off-street parking by a wide margin? Finally, how will CAVs impact public transportation? Will they help solve the first/last mile problem, or will they replace transit with their faster door-to-door service? In a CAV society, will we still need transit – and in what ways would these vehicles make cities more or less sustainable?

Adaptation to climate change

With each passing year, it is becoming more urgent to develop robust climate change adaptation strategies for transportation systems. The most recent National Climate Assessment (Schwartz et al., 2014) provides a useful framework for this task, identifying two forms of adaptation (Figure 3). The first aims to reduce, to the extent possible, the impacts of climate change on transport systems, mostly by infrastructure investment. The second aims to reduce the consequences of unavoidable impacts, mostly through strategic resilience planning for transport system disruption.

Equally important for cities is the effect of climate change on travel behavior. There is a growing literature on how weather affects travel choices, but many of the findings in one climate or culture are contradicted in another (Böcker, Dijst, & Prillwitz, 2013). Climate changes are likely to be felt most keenly by travelers exposed to the weather, which are precisely those using the modes that are considered more sustainable: walking, biking, and mass transit.

Concluding remarks

Substantial progress has been made in advancing urban transport sustainability. AFVs are finally out of the laboratory and on the streets. Built environment concepts such as TOD and complete streets have become mainstream. Cities across the globe are implementing Vision Zero policies to reduce traffic deaths. Urban air quality in the U.S. is far better than a generation ago. Despite these gains, GHGs from transportation are the fastest growing of any energy use, congestion continues to grow, and transport equity, affordability, safety, and health challenges remain.

If we agree that cars with current technology and fuels are a big part of the problem, should we invest our limited resources more heavily in making cars more sustainable or in getting people out of cars? The literature recognizes many co-benefits of sustainable travel behavior and land use solutions, but can we realistically expect societal behavior change to reduce car usage sufficiently and rapidly enough to address the challenge of climate change and reach our sustainability goals? Ultimately, this is the $64 billion (trillion?) question.

Organizations such as C40 Cities (2017, p. 1) tout that "cities have the power to change the world." Cities have the ability to implement bolder strategies faster than states, provinces, or countries, with their more diverse stakeholders (Rosenzweig, Solecki, Hammer, & Mehrotra, 2010). Preemption of urban sustainability initiatives by higher-level governments, however, has emerged as a political and legal battleground between cities, states, and the federal government (Burger, 2009; Graham, 2017).

Cities have nearly full control of their land use, but less control over their transport systems. An important takeaway from this paper is the overriding importance of integrated transportation and land use planning. This is not a new conclusion. Tight integration of land-use and transport systems has been an aspiration for 60 years (e.g., Clark, 1958; Wheeler, 1967). Some progress has been made (California's SB 375), but true solutions are elusive.

After a half century of mostly incremental change in our motorized transportation systems, we are now in the midst of a "disruption" or "transport revolution" (Gilbert & Perl, 2013), the implications of which will engage scholars of sustainable urban transportation for the (un)foreseeable future. It is worth remembering, however, that concerns about the sustainability of urban transportation once focused on depletion of grazing commons and emissions of horse manure. The quest for mobility has led humanity from non-human animate energy to motorized vehicles and inanimate energy, but each technological era has had its own issues with material throughput from finite resources to limited environmental capacity to absorb waste, and its own problems with spatial reorganization and economic, equity, and health consequences. If and when we succeed in replacing the automobile, it would be naïve to think that our urban transportation system will finally be environmentally, economically, and socially sustainable.

Disclosure statement

No potential conflict of interest was reported by the authors.

Funding

This work was supported by the University of North Carolina at Greensboro [FFA 2016].

ORCID

Selima Sultana http://orcid.org/0000-0003-4288-2518
Deborah Salon http://orcid.org/0000-0002-2240-8408
Michael Kuby http://orcid.org/0000-0002-7988-5766

References

Adams, Richard. (1992). Is happiness a home in the suburbs? The influence of urban versus suburban neighborhoods on psychological health. *Journal of Community Psychology*, 20(4), 353–372.
Agnolucci, Paolo, & McDowall, William. (2013). Designing future hydrogen infrastructure: Insights from analysis at different spatial scales. *International Journal of Hydrogen Energy*, 38(13), 5181–5191.
Alonso, Andre, Monzón, Andrés, & Cascajo, Rocío. (2015). Comparative analysis of passenger transport sustainability in European cities. *Ecological Indicators*, 48, 578–592.
Alonso, William. (1964). *Location and land use. Toward a general theory of land rent*. Cambridge: Harvard University Press.
Alternative Fuels Data Center. (2017). Federal and state laws and incentives. Retrieved from http://www.afdc.energy.gov/laws
Anderson, James, Kalra, Nidihi, Stanley, Karlyn, Sorensen, Paul, Samaras, Constantine, & Oluwatola, Tobi A. (2016). *Autonomous vehicle technology: A guide for policymakers*. Santa Monica, CA: RAND Corporation. Retrieved from https://www.rand.org/pubs/research_reports/RR443-2.html
Bäckström, Peter, Sandow, Erica, & Westerlund, Olle. (2016). Commuting and timing of retirement. *The Annals of Regional Science*, 56(1), 125–152.
Bergstad, Cecilia Jakobsson, Gamble, Amelie, Gärling, Tommy, Hagman, Olle, Polk, Merritt, Ettema, Dick, ... Olsson, Lars E. (2011). Subjective well-being related to satisfaction with daily travel. *Transportation*, 38(1), 1–15.
Berrigan, David, Pickle, Linda, & Dill, Jennifer. (2010). Associations between street connectivity and active transportation. *International Journal of Health Geographics*, 9(1), 20–37.
Bertolini, Luca, Le Clercq, Frank, & Kapoen, Loek. (2005). Sustainable accessibility: A conceptual framework to integrate transport and land use plan-making. Two test-applications in the Netherlands and a reflection on the way forward. *Transport Policy*, 12(3), 207–220.
Black, William. (1997a). The NSF-ESF research program on social change and sustainable transport: The Strasbourg papers. *Journal of Transport Geography*, 5(3), 163.
Black, William. (1997b). North American transportation: Perspectives on research needs and sustainable transportation. *Journal of Transport Geography*, 5(1), 12–19.
Black, William. (2010). *Sustainable transportation: Problems and solutions*. New York, NY: Guilford Press.
Boarnet, Marlon. (2011). A broader context for land use and travel behavior, and a research agenda. *Journal of the American Planning Association*, 77(3), 197–213.
Boarnet, Marlon, Houston, Douglas, Ferguson, Gavin, & Spears, Steven. (2011). Chapter 7: Land use and vehicle miles of travel in the climate change debate: Getting smarter than your average

bear. In G.K. Ingram & Y-H Hong (Eds.), *Climate change and land policies* (pp. 151–187). Cambridge, MA: Lincoln Institute for Land Policy. Retrieved from http://www.lincolninst.edu/sites/default/files/pubfiles/2038_1360_LP2010-ch07-Land-Use-and-Vehicle-Miles-of-Travel-in-the-Climate-Change-Debate_0.pdf

Böcker, Lars, Dijst, Martin, & Prillwitz, Jan. (2013). Impact of everyday weather on individual daily travel behaviours in perspective: A literature review. *Transport Reviews, 33*(1), 71–91.

British Petroleum. (2017). BP statistical review of world energy. Retrieved from http://www.bp.com/en/global/corporate/energy-economics/statistical-review-of-world-energy.html.

Brooks, Leah, & Lutz, Byron. (2016). From today's city to tomorrow's city: An empirical investigation of urban land assembly. *American Economic Journal: Economic Policy, 8*(3), 69–105.

Brown, Barbara B., Smith, Ken R., Tharp, Doug, Werner, Carol M., Tribby, Calvin P., Miller, Harvey J., & Jensen, Wyatt. (2016). A complete street intervention for walking to transit, nontransit walking, and bicycling: A quasi-experimental demonstration of increased use. *Journal of Physical Activity and Health, 13*(11), 1210–1219.

Brundtland Commission. (1987). *Our common future: Report of the world commission on environment and development* (UN Documents Gathering Body of Global Agreements), Oxford: Oxford University Press.

Brustman, Richard. (1999). *An analysis of available bicycle and pedestrian accident data: A report to the New York Bicycling Coalition*. Albany, NY: New York Bicycling Coalition.

Buehler, Ralph, & Pucher, John. (2017). Trends in walking and cycling safety: Recent evidence from high-income countries, with a focus on the United States and Germany. *American Journal of Public Health, 107*(2), 281–287.

Bunch, David, Bradley, Mark, Golob, Thomas, Kitamura, Ryuichi, & Occhiuzzo, Gareth. (1993). Demand for clean-fuel vehicles in California: A discrete-choice stated preference pilot project. *Transportation Research Part A: Policy and Practice*, Special Issue Energy and Global Climate Change, *27*(3), 237–253.

Bureau of Labor Statistics (BLS) U.S. Department of Labor. (2015). *2015 Consumer expenditure survey summary tables, Washington, DC: U.S. Department of Labor*.

Burger, Michael. (2009). It's not easy being green: Local initiatives, preemption problems, and the market participant exception. *University of Cincinnati Law Review, 78*, 835–889.

Busse, Meghan, Knittel, Christopher, & Zettelmeyer, Florian. (2013). Are consumers myopic? Evidence from new and used car purchases. *American Economic Review, 103*(1), 220–256.

C40Cities. (2017). Why cities? Ending climate change begins in the city. Retrieved from http://www.c40.org/why_cities

Caiazzo, Fabio, Ashok, Akshay, Waitz, Ian A., Yim, Steve, & Barrett, Steven. (2013). Air pollution and early deaths in the United States. Part I: Quantifying the impact of major sectors in 2005. *Atmospheric Environment, 79*, 198–208.

Cambridge Systematics. (2005). *Traffic congestion and reliability: Trends and advanced strategies for congestion mitigation* (Final Report). Texas Transportation Institute. Retrieved from http://ops.fhwa.dot.gov/congestion_report_04/index.html

Campbell, Scott. (1996). Green cities, growing cities, just cities?: Urban planning and the contradictions of sustainable development. *Journal of the American Planning Association, 62*(3), 296–312.

Campbell, Scott. (2016). The planner's triangle revisited: Sustainability and the evolution of a planning ideal that can't stand still. *Journal of the American Planning Association, 82*(4), 388–397.

Carlson, Susan A., Fulton, Janet E., Pratt, Michael, Yang, Zhou, & Adams, Kathleen E. (2015). Inadequate physical activity and health care expenditures in the United States. *Progress in Cardiovascular Diseases, 57*(4), 315–323.

Carroll, J.Douglass, & Bovls, Howard. (1957). Predicting local travel in urban regions. *Papers in Regional Science, 3*(1), 183–197.

Centre for Economics and Business Research. (2014). *The future economic and environmental costs of gridlock in 2030* (Report for INRIX). London. Retrieved from https://www.cebr.com/wp-content/uploads/2015/08/INRIX_costs-of-congestion_Cebr-report_v5_FINAL.pdf

Cervero, Robert, & Kockelman, Kara. (1997). Travel demand and the 3Ds: Density, diversity, and design. *Transportation Research Part D: Transport and Environment, 2*(3), 199–219.

Cervero, Robert, Murphy, Steven, Ferrell, Christopher, Goguts, Natasha, Tsai, Yu-Hsin, Arrington, G.B., & Witenstein, Nicole. (2004). *Transit-oriented development in the United States: Experiences, challenges, and prospects* (Transit Cooperative Research Program (TCRP) Report 102). Washington, DC: Transportation Research Board. Retrieved from http://onlinepubs.trb.org/onlinepubs/tcrp/tcrp_rpt_102.pdf

Chapman, Lee. (2007). Transport and climate change: A review. *Journal of Transport Geography, 15*(5), 354–367.

Chertow, Marian. (2000). The IPAT equation and its variants. *Journal of Industrial Ecology, 4*(4), 13–29.

Chester, Mikhail, Fraser, Andrew, Matute, Juan, Flower, Carolyn, & Pendyala, Ram. (2015). Parking infrastructure: A constraint on or opportunity for urban redevelopment? A study of Los Angeles County parking supply and growth. *Journal of the American Planning Association, 81*(4), 268–286.

Clark, Colin. (1958). Transport-maker and breaker of cities. *Town Planning Review, 28*(4), 237.

Curtis, Carey. (2008). Planning for sustainable accessibility: The implementation challenge. *Transport Policy, 15*(2), 104–112.

Curtis, Carey, & Low, Nicholas. (2016). *Institutional barriers to sustainable transport*. New York, NY: Routledge.

Dickinson, Robert. (1949). The geography of commuting in West Germany. *Annals of the Association of American Geographers, 49*(4), 443–456.

Dill, Jennifer, & Carr, Theresa. (2003). Bicycle commuting and facilities in major US cities: If you build them, commuters will use them. *Transportation Research Record: Journal of the Transportation Research Board, 1828*, 116–123.

Ding, Ding, Lawson, Kenny D., Kolbe-Alexander, Tracy L., Finkelstein, Eric A., Katzmarzyk, Peter T., Van Mechelen, Willem, & Pratt, Michael. (2016). The economic burden of physical inactivity: A global analysis of major non-communicable diseases. *The Lancet, 6736*(16), 30383–X. Retrieved from http://www.thelancet.com/pdfs/journals/lancet/PIIS0140-6736(16)30383-X.pdf

Downs, Anthony. (1992). *Stuck in traffic: Coping with peak-hour traffic congestion*. Washington, DC: The Brookings Institution. Cambridge, MA: The Lincoln Institute for Land Policy.

Downs, Anthony. (2004). Why traffic congestion is here to stay…. and will get worse. *ACCESS Magazine, 1*(25), 19-25.

Downs, Anthony. (2005). *Still stuck in traffic: Coping with peak-hour traffic congestion*. Washington, DC: The Brookings Institution.

Edwards, Robert, Mahieu, Vincent, Griesemann, Jean-Claude, Larivé, Jean-Francois, & Rickeard, David. (2004). *Well-to-wheels analysis of future automotive fuels and powertrains in the European context* (No. 2004-01-1924). SAE Technical Paper, Warrendale, PA.

El-Geneidy, Ahmed, Levinson, David, Diab, Ehab, Boisjoly, Genevieve, Verbich, David, & Loong, Charis. (2016). The cost of equity: Assessing transit accessibility and social disparity using total travel cost. *Transportation Research Part A: Policy and Practice, 91*, 302–316.

Ewing, Reid, & Cervero, Robert. (2010). Travel and the built environment. *Journal of the American Planning Association, 76*(3), 265–294.

Fan, Yueyue, Lee, Allen, Parker, Nathan, Scheitrum, Daniel, Jaffe, Amy M., Dominguez-Faus, Rosa, & Medlock, Kenneth, III. (2017). Geospatial, temporal and economic analysis of alternative fuel infrastructure: The case of freight and US natural gas markets. *Energy Journal, 38*(6), 199–230.

Farrell, Alexander, Keith, David, & Corbett, James. (2003). A strategy for introducing hydrogen into transportation. *Energy Policy, 31*(13), 1357–1367.

Friman, Margareta, Fujii, Satoshi, Ettema, Dick, Gärling, Tommy, & Olsson, Lars E. (2013). Psychometric analysis of the satisfaction with travel scale. *Transportation Research Part A: Policy and Practice, 48*, 132–145.

Galea, Sandro, Ahern, Jennifer, Rudenstine, Sasha, Wallace, Zachary, & Vlahov, David. (2005). Urban built environment and depression: A multilevel analysis. *Journal of Epidemiology and Community Health, 59*(10), 822–827.

Gilbert, Richard, & Perl, Anthony. (2013). *Transport revolutions: Moving people and freight without oil*. Gabriola Island, BC, Canada: New Society Publishers.

Gillingham, Kenneth. (2014). Identifying the elasticity of driving: Evidence from a gasoline price shock in California. *Regional Science and Urban Economics, 47*, 13–24.

Giuliano, Genevieve. (2005). Low income, public transit, and mobility. *Transportation Research Record: Journal of the Transportation Research Board, 1927*, 63–70.

Giuliano, Genevieve. (2012). Chapter 25: Transportation policy: Public transit, settlement patterns, and equity in the United States. In Nancy Brooks, Kieran Donaghy, & Gerrit-Jan Knaap (Eds.), *The Oxford handbook of urban economics and planning* (pp. 562–579). New York, NY: Oxford University Press.

Godish, Thad, Davis, Wayne, & Fu, Joshua. (2014). *Air quality*. Boca Raton, FL: CRC Press.

Golub, Aaron, Marcantonio, Richard, & Sanchez, Thomas. (2013). Race, space, and struggles for mobility: Transportation impacts on African Americans in Oakland and the East Bay. *Urban Geography, 34*(5), 699–728.

Goodwin, Phil, Dargay, Joyce, & Hanly, Mark. (2004). Elasticities of road traffic and fuel consumption with respect to price and income: A review. *Transport Reviews, 24*(3), 275–292.

Graham, Daniel, & Glaister, Stephen. (2004). Road traffic demand elasticity estimates: A review. *Transport Reviews, 24*(3), 261–274.

Graham, David. (2017). Red state, blue city. *The Atlantic, 319*(2), 24–26.

Greene, David. (1998). Why CAFE worked. *Energy Policy, 26*(8), 595–613.

Greene, David, Leiby, Paul, & Bowman, David. (2007). *Integrated analysis of market transformation scenarios with HyTrans* (ORNL/TM-2007/094). Oak Ridge, TN: Oak Ridge National Laboratory.

Grengs, Joe. (2002). Community-based planning as a source of political change: The transit equity movement of Los Angeles' Bus Riders Union. *Journal of the American Planning Association, 68*(2), 165–178.

Gudmundsson, Henrik, Hall, Ralph, Marsden, Greg, & Zietsman, Josias. (2016). *Sustainable transportation: Indicators, frameworks, and performance management*. Heidelberg, Germany: Springer.

Guo, Zhan, Agrawal, Asha W., & Dill, Jennifer. (2011). Are land use planning and congestion pricing mutually supportive? Evidence from a pilot mileage fee program in Portland, OR. *Journal of the American Planning Association, 77*(3), 232–250.

Guo, Zhan, & Ren, Shuai. (2013). From minimum to maximum: Impact of the London parking reform on residential parking supply from 2004 to 2010? *Urban Studies, 50*(6), 1183–1200.

Haas, P. M., Newmark, G. L., & Morrison, T. R. (2016). Untangling housing cost and transportation interactions: The location affordability index model—Version 2 (LAIM2). *Housing Policy Debate, 26*(4–5), 568–582. doi:10.1080/10511482.2016.1158199

Hirt, Sonia. (2016). The city sustainable: Three thoughts on "Green cities, growing cities, just cities.". *Journal of the American Planning Association, 82*(4), 383–384.

Huang, Yongxi, Kuby, Michael, & Chow, Joseph. (2017). Editorial for the virtual special issue on "Advances in alternative fuel vehicle transportation systems. *Transportation Research Part C: Emerging Technologies, 74*, 97–98.

Insurance Institute for Highway Safety Highway Loss Data Institute (IIHSHLDI). (2016). Pedestrians and bicyclists: Roadway improvements have been shown to reduce crashes. Retrieved from http://www.iihs.org/iihs/topics/t/pedestrians-and-bicyclists/fatalityfacts/bicycles

Kane, Kevin, York, Abigal, Tuccillo, Joseph, Gentile, Lauren, & Ouyang, Yun. (2014). Residential development during the Great Recession: A shifting focus in Phoenix, Arizona. *Urban Geography*, 35(4), 486–507.

Karner, Alexander, Eisinger, Douglas, & Niemeier, Deb. (2010). Near-roadway air quality: Synthesizing the findings from real-world data. *Environmental Science & Technology*, 44(14), 5334–5344.

Karner, Alexander, Kuby, Michael, & Golub, Aaron. (2017). *Modeling route-level transit ridership demographics for use in equity analysis*. Submitted for publication.

Kawabata, Mizuki, & Shen, Qing. (2007). Commuting inequality between cars and public transit: The case of the San Francisco Bay Area, 1990-2000. *Urban Studies*, 44(9), 1759–1780.

Kelley, Scott, & Kuby, Michael. (2013). On the way or around the corner? Observed refueling choices of alternative-fuel drivers in Southern California. *Journal of Transport Geography*, 33, 258–267.

Kenworthy, Jeffrey, & Laube, Felix. (1999). Patterns of automobile dependence in cities: An international overview of key physical and economic dimensions with some implications for urban policy. *Transportation Research Part A: Policy and Practice*, 33(7), 691–723.

Kidd, Charles. (1992). The evolution of sustainability. *Journal of Agricultural and Environmental Ethics*, 5(1), 1–26.

Kim, Hyojin. (2016). High-speed rail. In Selima Sultana & Joe Weber (Eds.), *Minicars, maglevs, and mopeds: Modern modes of transportation around the world: Modern modes of transportation around the world* (pp. 136–139). Santa Barbara: ABC-CLIO.

Kim, Hyojin, & Sultana, Selima. (2017). *A geographic assessment of the performance of high-speed rail stations on urban development: The implications from the Korea Train eXpress (KTX)*. Manuscript submitted for publication.

Kitamura, Ryuichi, Mokhtarian, Patricia, & Laidet, Laura. (1997). A micro-analysis of land use and travel in five neighborhoods in the San Francisco Bay Area. *Transportation*, 24(2), 125–158.

Koschinsky, Julia, & Talen, Emily. (2016). Location efficiency and affordability: A national analysis of walkable access and HUD-assisted housing. *Housing Policy Debate*, 26(4–5), 835–863.

Kuby, Michael, & Lim, Seow. (2005). The flow-refueling location problem for alternative-fuel vehicles. *Socio-Economic Planning Sciences*, 39(2), 125–145.

Lane, Bradley. (2010). The relationship between recent gasoline price fluctuations and transit ridership in major US cities. *Journal of Transport Geography*, 18(2), 214–225.

Lee, Bumsoo, & Lee, Yongsung. (2013). Complementary pricing and land use policies: does it lead to higher transit use? *Journal of the American Planning Association*, 79(4), 314–328.

Li, L., & Loo, B. P. (2016). Railway development and air patronage in China, 1993–2012: Implications for low-carbon transport. *Journal of Regional Science*. 57(3), 507–522.

Li, Ziru, Hong, Yili, & Zhang, Zhongju (2016). Do ride-sharing services affect traffic congestion? An empirical study of uber entry. Retrieved from SSRN https://ssrn.com/abstract=2838043

Litman, Todd (2016). Evaluating transportation equity: Guidance for incorporating distributional impacts in transportation planning. Retrieved from http://www.vtpi.org/equity.pdf

Litman, Todd, & Burwell, David. (2006). Issues in sustainable transportation. *International Journal of Global Environmental Issues*, 6(4), 331–347.

Lovejoy, Kristin, Sciara, Gian-Claudia, Salon, Deborah, Handy, Susan L., & Mokhtarian, Patricia. (2013). Measuring the impacts of local land-use policies on vehicle miles of travel: The case of the first big-box store in Davis, California. *Journal of Transport and Land Use*, 6(1), 25–39.

Lu, Chao, Rong, Ke, You, Jianxin, & Shi, Yongjiang. (2014). Business ecosystem and stakeholders' role transformation: Evidence from Chinese emerging electric vehicle industry. *Expert Systems with Applications*, 41(10), 4579–4595.

Lucas, Karen. (2012, March 31). Transport and social exclusion: Where are we now? *Transport Policy*, 20, 105–113.

Lucas, Karen, Van Wee, Bert, & Maat, Kess. (2016). A method to evaluate equitable accessibility: Combining ethical theories and accessibility-based approaches. *Transportation, 43*(3), 473–490.

Manaugh, Kevin, Badami, Madhav, & El-Geneidy, Ahmed. (2015). Integrating social equity into urban transportation planning: A critical evaluation of equity objectives and measures in transportation plans in North America. *Transport Policy, 37*, 167–176.

Manaugh, Kevin, & El-Geneidy, Ahmed. (2012). Chapter 12: Who benefits from new transportation infrastructure? Using accessibility measures to evaluate social equity in public transport provision. In Karst T. Geurs, Kevin J. Krizek, & Aura Reggiani (Eds.), *Accessibility analysis and transport planning: Challenges for Europe and North America*. Massachusetts: Edward Elgar Publishing.

Mann, Charles. (2013). What if we never run out of oil? *Atlantic Monthly, 311*(4), 48–63.

Marble, Duane. (1959). Transport inputs at urban residential sites. *Papers in Regional Science, 5*(1), 253–266.

Martin, Elliot, Shaheen, Susan, & Lidicker, Jeffrey. (2010). Impact of carsharing on household vehicle holdings: Results from North American shared-use vehicle survey. *Transportation Research Record: Journal of the Transportation Research Board, 2143*, 150–158.

Mattingly, Kerry, & Morrissey, John. (2014). Housing and transport expenditure: Socio-spatial indicators of affordability in Auckland. *Cities, 38*, 69–83.

May, Anthony, Kelly, Charlotte, & Shepherd, Simon. (2006). The principles of integration in urban transport strategies. *Transport Policy, 13*(4), 319–327.

McConnell, Virginia, & Wiley, Keith. (2010). *Infill development: Perspectives and evidence from economics and planning* (Discussion Paper 10-13). Washington, DC: Resources for the Future. Retrieved from http://www.rff.org/files/sharepoint/WorkImages/Download/RFF-DP-10-13.pdf

McFarlane, Colin. (2016). The geographies of urban density: Topology, politics and the city. *Progress in Human Geography, 40*(5), 629–648.

McLaughlin, Brian, Murrell, Scott, & DesRoches, Susanne. (2011). *Case study: Assessment of the vulnerability of Port Authority of NY & NJ facilities to the impacts of climate change* (TD&I Congress 2011: Integrated transportation and development for a better tomorrow). Reston, VA: American Society of Civil Engineers. Retrieved from http://ascelibrary.org/doi/abs/10.1061/41167%28398%2992

Meadows, Donella H, Meadows, Dennis L., Randers, Jørgen, & Behrens, William. (1972). *The limits to growth: A report to the Club of Rome (1972)*. New York, NY: Universe Books.

Melaina, Marc, & Bremson, Joel. (2008). Refueling availability for alternative fuel vehicle markets: Sufficient urban station coverage. *Energy Policy, 36*(8), 3233–3241.

Melendez, Margo. (2006). *Transitioning to a hydrogen future: Learning from the alternative fuels experience* (National Renewable Energy Laboratory). Retrieved from http://www.afdc.energy.gov/pdfs/39423.pdf

Mohammad, Sara, Graham, Daniel, Melo, Patricia, & Anderson, Richard. (2013). A meta-analysis of the impact of rail projects on land and property values. *Transportation Research Part A: Policy and Practice, 50*, 158–170.

Moroni, Stefano. (2016). Urban density after Jane Jacobs: The crucial role of diversity and emergence. *City, Territory and Architecture, 3*(1), 13.

National Highway Traffic and Safety Administration (NHTSA), United States Department of Transportation. (2017). Rural/urban comparison of traffic fatalities. Retrieved from https://crashstats.nhtsa.dot.gov/

National Household Travel Survey. (2009). Retrieved from http://nhts.ornl.gov/

National Research Council (US). (2008). *Potential impacts of climate change on US transportation* (Transportation Research Board Special Report 290). Washington, DC: National Academy. Retrieved from http://onlinepubs.trb.org/onlinepubs/sr/sr290.pdf.

Nesbitt, Kevin, & Sperling, Daniel. (1998). Myths regarding alternative fuel vehicle demand by light-duty vehicle fleets. *Transportation Research Part D: Transport and Environment, 3*(4), 259–269.

Ogden, Joan, Fulton, Lew, & Sperling, Daniel. (2016). *Making the transition to light-duty electric-drive vehicles in the US: Costs in perspective to 2035* (Research Report – UCD-ITS-RR-16-21). Davis, CA: University of California at Davis, Institute for Transportation Studies. Retrieved from https://www.globalfueleconomy.org/media/418729/the-relative-cost-of-transport-transitions-ogden-fulton.pdf

Ogden, Joan, & Nicholas, Michael. (2011). Analysis of a "cluster" strategy for introducing hydrogen vehicles in Southern California. *Energy Policy, 39*(4), 1923–1938.

Ottosson, Dadi B, Chen, Cynthia, Wang, Tingting, & Lin, Haiyun. (2013). The sensitivity of on-street parking demand in response to price changes: A case study in Seattle, WA. *Transport Policy, 25*(Jan), 222–232.

Parkany, Emily, Gallagher, Ryan, & Viveiros, Phillip. (2004). Are attitudes important in travel choice? *Transportation Research Record: Journal of the Transportation Research Board, 1894*, 127–139.

Parry, Ian, Walls, Margaret, & Harrington, Winston. (2007). Automobile externalities and policies. *Journal of Economic Literature, 45*(2), 373–399.

Pearre, Nathaniel, Kempton, Willett, Guensler, Randal, & Elango, Vetri. (2011). Electric vehicles: How much range is required for a day's driving? *Transportation Research Part C: Emerging Technologies, 19*(6), 1171–1184.

Pereira, Rafael, Schwanen, Tim, & Banister, David. (2017). Distributive justice and equity in transportation. *Transport Reviews, 37*(2), 170–191.

Pfeiffer, Deirdre, & Cloutier, Scott. (2016). Planning for happy neighborhoods. *Journal of the American Planning Association, 82*(3), 267–279.

Piatkowski, Daniel P., Krizek, Kevin J., & Handy, Susan. (2015). Accounting for the short term substitution effects of walking and cycling in sustainable transportation. *Travel Behaviour and Society, 2*(1), 32–41.

Pierce, Gregory, & Shoup, Donald. (2013). Getting the prices right: An evaluation of pricing parking by demand in San Francisco. *Journal of the American Planning Association, 79*(1), 67–81.

Pollack, Stephanie, Bluestone, Barry, & Billingham, Chase. (2010). *Maintaining diversity in America's transit rich neighborhoods: Tools for equitable neighborhood change*. Dukakis Center for Urban and Regional Policy, Boston: MA. Retrieved from http://www.reconnectingamerica.org/assets/Uploads/TRNEquityfinal.pdf

Potoglou, Dimitris, & Kanaroglou, Pavlos. (2007). An internet-based stated choices household survey for alternative fueled vehicles. *Canadian Journal of Transportation, 1*(1), 36–55.

Pratt, Edward E. (1911). *Industrial causes of congestion of population in New York City* (No. 109-111). NewYork: Columbia University, Longmans, Green & Company.

Pratt, Michael, Norris, Jeffrey, Lobelo, Felipe, Roux, Larissa, & Wang, Guijing. (2014). The cost of physical inactivity: Moving into the 21st century. *British Journal of Sports Medicine, 48*(3), 171–173.

Preston, John. (2009). Epilogue: Transport policy and social exclusion—Some reflections. *Transport Policy, 16*(3), 140–142.

Pucher, John, Dill, Jennifer, & Handy, Susan. (2010). Infrastructure, programs, and policies to increase bicycling: An international review. *Preventive Medicine, 50*(S), 106–S125.

Puentes, Robert, & Roberto, Elizabeth. (2008). *Commuting to opportunity: The working poor and commuting in the United States* (Report). Washington, DC: The Brookings Institution.

Purvis, Martin, & Grainger, Alan. (2004). *Exploring sustainable development: Geographical perspectives*. New York, NY: Routledge.

Pyrialakou, Dimitra, Gkritza, Konstantin, & Fricker, Jon. (2016). Accessibility, mobility, and realized travel behavior: Assessing transport disadvantage from a policy perspective. *Journal of Transport Geography, 51*, 252–269.

Randall, Todd, & Baetz, Brian. (2001). Evaluating pedestrian connectivity for suburban sustainability. *Journal of Urban Planning and Development, 127*(1), 1–15.

Rayle, Lisa, Dai, Danielle, Chan, Nelson, Cervero, Robert, & Shaheen, Susan. (2016). Just a better taxi? A survey-based comparison of taxis, transit, and ridesourcing services in San Francisco. *Transport Policy, 45*(January), 168–178.

Rosenzweig, Cynthia, Solecki, William, Hammer, Stephen A., & Mehrotra, Shagun. (2010). Cities lead the way in climate-change action. *Nature, 467*(7318), 909–911.

Saelens, Brian, & Handy, Susan. (2008). Built environment correlates of walking: A review. *Medicine and Science in Sports and Exercise, 40*(7 Suppl), S550.

Sallis, James F., Cerin, Ester, Conway, Terry L., Adams, Marc A., Frank, Lawrence D., Pratt, Michael, & Owen, Neville. (2016). Physical activity in relation to urban environments in 14 cities worldwide: A cross-sectional study. *The Lancet, 387*(10034), 2207–2217.

Salon, Deborah. (2015). Heterogeneity in the relationship between the built environment and driving: Focus on neighborhood type and travel purpose. *Research in Transportation Economics, 52*, 34–45.

Salon, Deborah, Boarnet, Marlon G., Handy, Susan, Spears, Steven, & Tal, Gil. (2012). How do local actions affect VMT? A critical review of the empirical evidence. *Transportation Research Part D: Transport and Environment, 17*(7), 495–508.

Salon, Deborah, & Gulyani, Sumila. (2010). Mobility, poverty, and gender: Travel 'choices' of slum residents in Nairobi, Kenya. *Transport Reviews, 30*(5), 641–657.

Schipper, Lee. (2011). Automobile use, fuel economy and CO 2 emissions in industrialized countries: Encouraging trends through 2008? *Transport Policy, 18*(2), 358–372.

Schwartz, Henry, Meyer, Michael, Burbank, Cynthia, Kuby, Michael, Oster, Clinton, Posey, John, & Rypinski, Arthur. (2014). Ch. 5: Transportation. In J. M. Melillo, Terese (T.C.) Richmond, & G. W. Yohe (Eds.), *Climate change impacts in the United States: The third national climate assessment* (pp. 130–149). Washington, DC: U.S. Global Change Research Program.

Smart Growth America. (2017). National complete streets coalition. Retrieved from https://smartgrowthamerica.org/program/national-complete-streets-coalition/

Spears, Steven, Boarnet, Marlon G., & Houston, Douglas. (2016). Driving reduction after the introduction of light rail transit: Evidence from an experimental-control group evaluation of the Los Angeles Expo Line. *Urban Studies.* 54(12), 2780-2799.

Spector, Julia, & Pyper, Julia. (2017). Updated: 17 states now charge fees for electric vehicles. Retrieved from https://www.greentechmedia.com/articles/read/13-states-now-charge-fees-for-electric-vehicles#gs.l1Zx9Xs

Sperling, Daniel, & Gordon, Deborah. (2009). *Two billion cars: Driving toward sustainability*. Oxford : Oxford University Press.

Sperling, Daniel, & Kitamura, Ryuichi. (1986). Refueling and new fuels: An exploratory analysis. *Transportation Research Part A: General, 20*(1), 15–23.

Stangl, Paul. (2012). The pedestrian route directness test: A new level-of-service model. *Urban Design International, 17*(3), 228–238.

Stradling, David, & Thorsheim, Peter. (1999). The smoke of great cities: British and American efforts to control air pollution, 1860-1914. *Environmental History, 4*(1), 6–31.

Struben, Jeroen, & Sterman, John. (2008). Transition challenges for alternative fuel vehicle and transportation systems. *Environment and Planning B: Planning and Design, 35*(6), 1070–1097.

Sultana, Selima. (2014). Commuting in America. In Mark Garrett (Eds.), *Encyclopedia of transportation: Social science and policy* (pp. 396–399). Thousand Oaks, CA: Sage Publication.

Sultana, Selima. (2015). Factors associated with students' parking-pass purchase decisions: Evidence from an American University. *Transport Policy, 44*, 65–75.

Sultana, Selima. (2017). Land Use and Transportation. In D. Richardson, N. Castree, M. F. Goodchild, A. L. Kobayashi, W. Liu, & R. A. Marston (Eds.), *The international encyclopedia of geography: People, the earth, environment and technology (IEG)* (pp. 1–11). Hoboken, NJ: John Wiley & Sons.

Sultana, Selima, & Weber, Joe. (2007). Journey-to-work patterns in the age of sprawl: Evidence from two mid–size southern metropolitan areas. *The Professional Geographer, 59*(2), 193–208.

Sultana, Selima, & Weber, Joe. (2014). The nature of urban growth and the commuting transition: Endless sprawl or a growth wave? *Urban Studies, 51*(3), 544–576.

Tal, Gil, & Handy, Susan. (2012). Measuring nonmotorized accessibility and connectivity in a robust pedestrian network. *Transportation Research Record: Journal of the Transportation Research Board*, *2299*, 48–56.

Taylor, Brian. (2017). The geography of urban transportation finance. In Susan Hanson & Genevieve Giuliano (Eds.), *The geography of urban transportation* (pp. 281–286). New York, NY: Guilford.

Taylor, Brian, Miller, Douglas, Iseki, Hiroyuki, & Fink, Camille. (2009). Nature and/or nurture? Analyzing the determinants of transit ridership across US urbanized areas. *Transportation Research Part A: Policy and Practice*, *43*(1), 60–77.

Taylor, Catherine, Hyde, David, & Barr, Lawrence. (2016). *Hyperloop commercial feasibility analysis: High level overview* (No. DOT-VNTSC-NASA-16-01). Retrieved from https://ntl.bts.gov/lib/59000/59300/59393/DOT-VNTSC-NASA-16-01.pdf

Taylor, Graham R. (1915). *Satellite cities: A study of industrial suburbs*. New York, NY: Appleton.

Terrence, Farris, J. (2001). The barriers to using urban infill development to achieve smart growth. *Housing Policy Debate*, *12*(1), 1–30.

Transit Oriented Development Institute. (2017). Transit oriented development. Retrieved from http://www.tod.org/

U. S. Office of Management and Budget (USOMB). (2016). *2016 Draft report to Congress on the benefits and costs of Federal regulations and agency compliance with the Unfunded Mandates Reform Act*. Washington, DC: Office of Management and Budget.

U.N. Habitat (Ed.). (2016). Urbanization and development: Emerging futures. Retrieved from http://nua.unhabitat.org/uploads/WCRFullReport2016_EN.pdf

U.S. Department of Energy (USDOE). (2016). Fact #948: October 24 2016 Carbon dioxide emissions from transportation exceeded those from the electric power sector for the first time in 38 years – dataset. Washington, DC: U.S. Department of Labor. Retrieved from https://energy.gov/eere/vehicles/downloads/fact-948-october-24-2016-carbon-dioxide-emissions-transportation-exceeded.

U.S. Environmental Protection Agency (USEPA). (2016). Smog, soot, and other air pollution from transportation. Retrieved from https://www.epa.gov/air-pollution-transportation/smog-soot-and-local-air-pollution

Van Acker, Veronique, Van Wee, Bert, & Witlox, Frank. (2010). When transport geography meets social psychology: Toward a conceptual model of travel behaviour. *Transport Reviews*, *30*(2), 219–240.

Van Wee, Bert, & Handy, Susan. (2016). Key research themes on urban space, scale, and sustainable urban mobility. *International Journal of Sustainable Transportation*, *10*(1), 18–24.

Veitch, J., Carver, A., Salmon, J., Abbott, G., Ball, K., Crawford, D., . . . Timperio, A. (2017). What predicts children's active transport and independent mobility in disadvantaged neighborhoods? *Health & Place*, *44*, 103–109.

Venner, Marie, & Ecola, Lisa. (2007). Financing transit-oriented development: Understanding and overcoming obstacles. *Transportation Research Record: Journal of the Transportation Research Board*, *1996*, 17–24.

Vickerman, Roger. (2015). High-speed rail and regional development: The case of intermediate stations. *Journal of Transport Geography*, *42*, 157–165.

Vision Zero. (2017). Vision zero. Retrieved from http://www1.nyc.gov/site/visionzero/initiatives/initiatives.page

Von Blottnitz, Harro, & Curran, Mary Ann. (2007). A review of assessments conducted on bio-ethanol as a transportation fuel from a net energy, greenhouse gas, and environmental life cycle perspective. *Journal of Cleaner Production*, *15*(7), 607–619.

Wachs, Martin. (2010). Transportation policy, poverty, and sustainability: History and future. *Transportation Research Record: The Journal of the Transportation Research Board*, *2163*, 3–12.

Wheeler, James. (1967). Occupational status and work-trips: A minimum distance approach. *Social Forces*, *45*(4), 508–515.

White, R. D. (2011, March 15). Regular gasoline is nearing $4 a gallon and could hit $5 a gallon. *Los Angeles Times*. Retrieved from http://articles.latimes.com/2011/mar/15/business/la-fi-gas-prices-20110315

World Health Organization. (2010). *Global recommendation on physical activity for health*. Geneva, Switzerland: WHO Press.

World Health Organization. (2016). Road traffic injuries. Retrieved from http://www.who.int/mediacentre/factsheets/fs358/en/

Wygonik, Erica, & Goodchild, Anne. (2016). Urban form and last-mile goods movement: Factors affecting vehicle miles travelled and emissions. *Transportation Research Part D: Transport and Environment*. Published online.

Xiong, Yubing, & Zhang, Junyi. (2016). Effects of land use and transport on young adults' quality of life. *Travel Behaviour and Society*, 5, 37–47.

Ye, Liang, Mokhtarian, Patricia L., & Circella, Giovanni. (2012). Commuter impacts and behavior changes during a temporary freeway closure: The 'Fix I-5'project in Sacramento, California. *Transportation Planning and Technology*, 35(3), 341–371.

Zavestoski, Stephen, & Agyeman, Julian (Eds.). (2014). *Incomplete streets: Processes, practices, and possibilities*.London: Routledge.

Zhang, Junping, Wang, Fei-Yue, Wang, Kunfeng, Lin, Wei-Hue, Xin, Xu, & Chen, Cheng. (2011). Data-driven intelligent transportation systems: A survey. *IEEE Transactions on Intelligent Transportation Systems*, 12(4), 1624–1639.

Urban resilience for whom, what, when, where, and why?

Sara Meerow and Joshua P. Newell

ABSTRACT
In academic and policy discourse, the concept of urban resilience is proliferating. Social theorists, especially human geographers, have rightfully criticized that the underlying politics of resilience have been ignored and stress the importance of asking "resilience of what, to what, and for whom?" This paper calls for careful consideration of not just resilience for whom and what, but also where, when, and why. A three-phase process is introduced to enable these "five Ws" to be negotiated collectively and to engender critical reflection on the politics of urban resilience as plans, initiatives, and projects are conceived, discussed, and implemented. Deployed through the hypothetical case of green infrastructure in Los Angeles, the paper concludes by illustrating how resilience planning trade-offs and decisions affect outcomes over space and time, often with significant implications for equity.

1. Introduction

Urbanization processes drive change in the Anthropocene, presenting environmental and social challenges that are unprecedented in scale, scope, and complexity (Seto, Sánchez-Rodríguez, & Fragkias, 2010). Climate change introduces additional uncertainties, placing pressure on local institutions to adapt. To marshal the actors and resources necessary for cities to effectively adjust and sustain key functions, academics and policymakers are turning to the concept of "urban resilience" as an organizing principle (Leichenko, 2011). In both the broader academy and public discourse, the concept's growing popularity is evident. Figure 1 illustrates the exponential increase in studies that apply the concept of resilience to cities, a trend especially pronounced in the fields of climate change and hazards (Beilin & Wilkinson, 2015). Policy initiatives related to urban resilience are also proliferating (Vale, 2014)[1].

One of the attractions of the resilience concept is its ability to serve as a "boundary object" (Brand & Jax, 2007) or "bridging concept" (Beichler, Hasibovic, Davidse, & Deppisch, 2014), thereby allowing multiple knowledge domains to interface. The shared concept of urban resilience, for example, has helped fuse the "climate change adaptation" and "disaster risk reduction" agendas (ARUP, 2014, p. 3), as well as security and sustainability priorities (Coaffee, 2008). But the term's flexibility and inherent inclusiveness has also led to conceptual confusion, especially in relation to

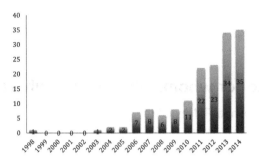

Figure 1. The rapid rise of urban resilience research: a graph showing the number of publications in the Web of Science database for each year from 1998 to 2014 with the terms "urban resilience," "resilient city," or "resilient cities" in the title, abstract, or keywords. *Note*: This may be an underestimate as Web of Science has stronger coverage of the natural sciences and engineering than social sciences.

like-minded terms such as sustainability, vulnerability, and adaptation (Elmqvist, 2014). These concepts are all commonly used in urban studies and policy, but in a multitude of ways, including as measurable characteristics, descriptive concepts, metaphors, and modes of thinking or paradigms.

Nevertheless, the theoretical roots of resilience give it a particular focus and connotation that makes a resilience approach related to, but distinct from, sustainability, adaptation, and vulnerability.[2] In the influential ecological and social-ecological systems (SES) resilience literature, systems thinking is pervasive. The focus in this work has traditionally been on quantitative modeling rather than the interactions between individual components and dynamics within the boundaries of a system (Turner, 2014). The most trenchant critiques of resilience scholarship come from social theorists, who take issue with the ways in which ecological models are applied to social structures and the general lack of attentiveness to issues of politics, power, and equity (Cote & Nightingale, 2011; Cretney, 2014; Evans, 2011; MacKinnon & Derickson, 2012; Weichselgartner & Kelman, 2015). These scholars rightfully assert the need to consider questions of "resilience of what to what?" and "resilience for whom?" (Carpenter, Walker, Anderies, & Abel, 2001; Lebel et al., 2006; Vale, 2014), as well as to reflect on scalar and temporal trade-offs (Chelleri, Waters, Olazabal, & Minucci, 2015).

Yet the popularity of resilience, especially in policy discourse, continues to grow. As Weichselgartner and Kelman (2015, p. 254) recognize, "While the academic debate on describing resilience continues, governments around the world have developed plans and programmes that aim to guide cities, communities and authorities towards achieving it."

In this paper, we argue that the resilience concept is redeemable. What is missing is a process by which to incorporate these important critiques. The primary objective of this paper, therefore, is to introduce such a process, which can be divided into three phases. The first involves the establishment of urban resilience as a boundary object, in which collaborators share a common definition of resilience and come to a basic agreement on what is "urban." The second phase entails critically thinking through resilience for whom, what, when, where, and why. These "five Ws of urban resilience" shape how resilience is operationalized and mapped over time and space. The third phase then explores urban resilience in empirical contexts. Taken together, this approach engenders a politics of resilience that includes grappling with trade-offs and scalar complexities and delineating how political context and power dynamics shape resilience policies, with inevitable winners and losers.

The next section briefly reviews the origins of the resilience concept and compares it with sustainability, vulnerability, and adaptation. Then Section 3 introduces the three-phase process designed to foster a politics of urban resilience, detailing in particular the five Ws. This is followed by Section 4, which uses a hypothetical example of green infrastructure planning for the city of Los Angeles to illustrate the ways in which questions of who, what, when, where, and why have wide-ranging implications for communities, institutions, and ecologies. The paper concludes by briefly considering how geographers could enrich urban resilience research.

2. The concept of resilience in the literature

Understanding the concept of urban resilience requires knowledge of how resilience theory has developed. Although the term has a long history of use in psychology and engineering, in the global environmental change literature, resilience is commonly traced back to ecologist C.S. Holling (Brown, 2014; Garschagen, 2013; Meerow & Newell, 2015). Holling (1973) defined resilience as an ecosystem's ability to maintain basic functional characteristics in the face of disturbance. Characterizing ecosystems as having multiple stable states and in a constant state of flux, Holling (1996) later distinguished between *static* "engineering" resilience, referring to a system's ability to bounce back to its previous state, and *dynamic* "ecological" resilience, which focuses on maintaining key functions when perturbed.

This ecological framing of resilience and understanding of ecosystems as dynamic, complex, and adaptive was seminal to the development of socio-ecological system (SES) theory, led by a group of interdisciplinary-minded ecologists (Folke, 2006; Gunderson & Holling, 2002). SES theory effectively extended Holling's ecological concepts to the "social" by conceptualizing nature-society as an intertwined, coevolving system. In the SES literature, resilience is identified as a product of (1) the amount of perturbation a system can endure without losing its key functions or changing states, (2) the system's ability to self-organize, and (3) the system's capacity for adaptation and learning (Folke et al., 2002).

The resilience concept has been applied in a wide range of empirical contexts, extending it from a descriptive term (i.e. reflecting how an ecosystem functions) to a normative approach or "way of thinking" (Folke, 2006, p. 260). This approach has become foundational for thinking through how complex systems can persist in the face of uncertainty, disruption, and change (Davoudi et al., 2012; Matyas & Pelling, 2014). Cities have been identified as the "example par excellence of complex systems" (Batty, 2008, p. 769); therefore, it is no surprise that resilience theory is increasingly applied in urban studies (Elmqvist, 2014; Leichenko, 2011; Meerow, Newell, & Stults, 2016). In its original, more descriptive form, resilience can be both positive and negative; however, "resilience thinking" and the concept of "resilient cities" have emerged as normative, desired goals in both academic and policy arenas (Cote & Nightingale, 2011; Vale, 2014). These different uses of the term have led to a multitude of definitions and confusion about what resilience means and how it relates to other key concepts like sustainability, vulnerability, and adaptation, which we turn to next.

2.1. *Parsing differences: resilience, sustainability, adaptation, vulnerability*

Conceptually, the relationship between resilience and sustainability is often muddled (Redman, 2014). Sustainability is usually linked to "sustainable development," defined in the Bruntland Report (Bruntland, 1987) as: "Development that meets the needs of the present without compromising the ability of future generations to meet their own needs". In some instances, sustainability and resilience are used interchangeably, in others resilience is presented as an important component of broader sustainability goals, and resilience has even been heralded as a new and improved paradigm (Derissen, Quaas, & Baumgärtner, 2011). Leading resilience scholars have generally argued that system resilience is crucial for achieving sustainability in "a world of transformations" (Folke et al., 2002). Thus, as a descriptive concept, resilience does not necessarily conflict with sustainability. Due to different theoretical legacies, however, when conceived as a way of thinking, or as a paradigm of environmental change and management, there are notable distinctions.

In the SES resilience literature, systems exist in a constant state of flux, requiring flexible planning and management (Folke, 2006). In comparison, some resilience thinkers find sustainability management approaches that seek an optimal balance between current and future needs problematically "static" (Cascio, 2009, p. 92). In other words, rather than predicting and planning for a more sustainable future, resilience stresses uncertainty and building systems-based adaptive capacity to unexpected future changes(Meerow & Baud, 2012). There are situations in which this conflicts with traditional sustainability goals. Sustainability measures often seek to optimize eco-efficiency, yet research suggests that functional redundancy fosters resilience (Korhonen & Seager, 2008). So, "an efficient optimal state outcome" (Walker & Salt, 2006, p. 9) could conceivably reduce resilience rather than foster it. Similarly, Redman (2014, p. 8) points out that so-called "smart cities" are often presented as more sustainable, yet the increased efficiency and interconnectedness of smart cities suggests "an inflexibility and extreme hypercoherence that resilience theorists have often warned against."

There are other important differences. Resilience emphasizes systems-based modeling and relies on SESs as the basic unit of analysis. This can obfuscate inequalities within the system, fail to account for the range of social actors involved, and pay insufficient attention to social dynamics (Bahadur & Tanner, 2014; Leach, 2008).[3] In the sustainability literature, there is a strong emphasis on balancing economic, environmental, and social justice goals (Brand & Jax, 2007). In the resilience scholarship, such concepts receive less attention (Friend & Moench, 2013).

Concern with social equity and political issues also distinguish the vulnerability and adaptation scholarship from the resilience literature. Although all three research domains share an interest in linked human-natural systems and how these SESs cope with disruptions and change, as Miller et al. (2010, p. 6) observe, adaptation and vulnerability research provides a "more politically nuanced understanding of social change and equity." In contrast to work on resilience, constructivist social scientists have heavily influenced the vulnerability and adaptation research (Miller et al., 2010). By focusing on studies of human actors and communities and how the environment poses a threat or provides resources to them, this research also tends to be more anthropocentric than resilience studies (Turner, 2010). While adaptation and vulnerability research is somewhat interconnected, resilience scholarship is more isolated (Janssen, Schoon, Ke, & Börner, 2006). Collaboration between these research communities may be undermined by conceptual confusion. In some instances, as with sustainability and resilience, the terms are used interchangeably. At other times, they are inversely related, with resilience seen as the flipside of vulnerability or even as one determinant of it (Gallopín, 2006).

2.2. Theoretical critiques of resilience

A number of geographers and social scientists contend that issues of power, scale, and equity are not given sufficient attention when considering the resilience of SESs (Cote & Nightingale, 2011; Cretney, 2014; Evans, 2011; MacKinnon & Derickson, 2012; Pizzo, 2015; Weichselgartner & Kelman, 2015). They are especially concerned with the ramifications of applying ecological models to society, as well as how resilience as a concept is deployed and by whom. In other words, "resilience of what, to what, and for whom?" (Elmqvist, 2014). As a whole, this emerging critical discourse focuses on three shortcomings: (1) a general lack of clarity with respect to meaning, (2) failure to sufficiently address scalar dimensions and trade-offs, and (3) inherent conservatism and the resulting preservation of the status quo.

The concept of resilience is commonly criticized for being too ambiguous and difficult to operationalize or measure (Matyas & Pelling, 2014; Vale, 2014). As resilience is adapted to a wide array of disciplines and policy sectors, there is concern that it may lose meaning and become an "empty signifier" (Weichselgartner & Kelman, 2015).

Depending on how resilience is operationalized, it can lead to spatial and temporal trade-offs and inequitable benefits, but these issues have not been sufficiently scrutinized (Chelleri et al., 2015). Part of the problem has to do with the transference of an ecological concept (i.e. resilient ecosystems) to social systems, at least initially by scholars not especially familiar with complexities associated with studying how society functions (Brown, 2014). For MacKinnon and Derickson (2012), resilience approaches oversimplify issues of spatial scale because they tend to view cities or communities as a "self-organizing" unit, akin to an ecosystem, that must protect itself from external threats. This artificially separates them from wider scales and processes. Conceptualizing cities as predictable or generalizable systems has also been criticized as a theoretical regression (Beilin & Wilkinson, 2015), ignoring decades of work on urban interconnectedness and inequality by urban theorists (e.g. see Brenner and Schmid, 2011, Harvey, 1996, and Heynen, Kaika, and Swyngedouw, 2006).

For Joseph (2013) and others, the resilience agenda is inherently conservative and tends to perpetuate an unjust status quo (Cretney, 2014; MacKinnon & Derickson, 2012; Walker & Cooper, 2011; Welsh, 2014). By assuming that complex systems naturally go through adaptive cycles of collapse and reorganization, ecological resilience theory "accepts change somewhat passively," often precluding the consideration of the social causes of crises (Evans, 2011, p. 224). The onus is placed on individuals or communities to adapt to inevitable disruptions, rather than addressing the underlying causes of these crises (Wamsler, 2014). For some, this resonates with neoliberal efforts to roll back the responsibilities of the state (Joseph, 2013; MacKinnon & Derickson, 2012; Welsh, 2014). As Evans and Reed (2014, p. 1) write, the resilience agenda is an effort on the part of liberal regimes to create a "catastrophic imaginary that promotes insecurity by design." Similarly, Walker and Cooper (2011) attribute the popularity of resilience theory to its ideological fit with the influential complexity theory-based financial system models of Friedrich Hayek.

For MacKinnon and Derickson (2012), a focus on resilience impedes necessary systemic transformation. Indeed, in analyzing the discourse of major international organizations' resilience-building initiatives, Brown (2012) found that resilience supported business as usual. In response, some leading resilience scholars have attempted to integrate transformation into resilience thinking, in addition to recovery and adaptability (see Olsson, Galaz, and Boonstra 2014 for a discussion). Nevertheless, MacKinnon and Derickson (2012) argue for replacing resilience with "resourcefulness," which they feel better supports social justice by providing marginalized communities with the capacity to transform society and enact their own desired futures.

While critical social scientists may ultimately disagree on the value of the resilience concept, together they highlight the need to examine the underlying politics of resilience. This includes questioning who sets the resilience agenda, how resilience is conceptualized, at what scales it is applied, and who benefits or loses.

3. Enabling a politics of urban resilience

This section introduces an iterative three-phase process to facilitate a politics of urban resilience in which knowledge is coproduced by decision makers and researchers and ideally leads to more usable science (Dilling & Lemos, 2011; Jasanoff, 2004) (Figure 2).

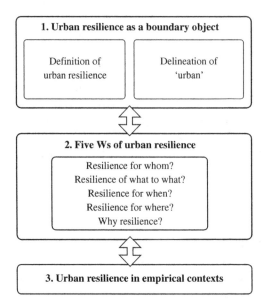

Figure 2. Process for enabling a politics of urban resilience.

Phase 1 involves conceptualizing urban resilience as a boundary object based on a shared definition and understanding of what is included in the "urban system." In phase 2, questions related to resilience for whom, what, where, when, and why are carefully considered. This forms the basis for testing, modeling, and applying urban resilience in empirical contexts (phase 3), thereby advancing both knowledge and practice.

3.1. Urban resilience as a boundary object

The concept of urban resilience serves a valuable function by initiating multidisciplinary dialogue; however, some consensus on both the meaning of "resilience" and "urban" provides a stronger basis for collaboration. Thus, in phase 1, an inclusive definition of urban resilience and conceptual schematic of the urban serve as a boundary object, bringing together different stakeholders and disciplines.

A boundary object refers to an object or concept that resonates with different social worlds, and as a result, supports scientific collaboration across disciplines (Star & Griesemer, 1989). A boundary object's meaning is somewhat flexible, which allows it to be adapted to the needs of various disciplines and stakeholders. Previous studies have shown that resilience effectively functions as a boundary object or bridging concept (Beichler et al., 2014; Brand & Jax, 2007; Coaffee, 2013). As Vale (2014, p. 198) argues, "the biggest upside to resilience, however, is the opportunity to turn its flexibility to full advantage by taking seriously the actual interconnections among various domains that have embraced the same terminology." While some malleability in the meaning of resilience may foster collaboration, too much ambiguity makes it difficult to operationalize resilience for any specific policy context (Matyas & Pelling, 2014).

Like the broader concept of resilience, urban resilience has become an increasingly popular, but also increasingly vague term (Meerow et al., 2016). This ambiguity hinders effective operationalization, benchmarking, and measurement of resilience (Pizzo, 2015). A shared interest in building more resilient cities may bring different disciplines to the table, but conceptual tensions have made consensus on a shared definition elusive (Beichler et al., 2014). Some agreement on a common definition of urban resilience is needed to avoid it becoming an empty signifier (Vale, 2014). Therefore, Meerow et al. (2016) recently proposed the following definition:

> Urban resilience refers to the ability of an urban system—and all its constituent socio-ecological and socio-technical networks across temporal and spatial scales—to maintain or rapidly return to desired functions in the face of a disturbance, to adapt to change, and to quickly transform systems that limit current or future adaptive capacity.

Part of what makes urban resilience so difficult to define is the inherent complexity of cities (Jabareen, 2013). Geographers and urban scholars have long debated what constitutes the "urban." Should cities be understood as individual bounded systems or even ecosystems (Pickett et al., 2001), as linked systems of cities (Ernstson et al., 2010), or a complex system of networks (Desouza & Flanery, 2013)? Developing a conceptual model of the urban requires delineating the various political, social, ecological, and technical features of cities as well as complex urban–rural and city-to-city linkages and resource flows. Figure 3 represents a conceptual model of an urban system developed by Meerow et al. (2016), which is composed of four interconnected components: (1)

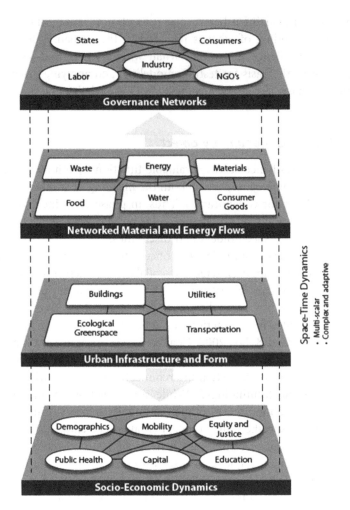

Figure 3. A conceptual schematic of the urban system proposed by Meerow et al. (2016) and inspired by Dicken (2011).

governance networks; (2) networked material and energy flows; (3) urban infrastructure and form; and (4) socioeconomic dynamics, all of which interact across spatial and temporal scales. A conceptual schematic like this one can help structure meaningful discussions about the complex and multiscalar components of cities or what is meant by "urban" in urban resilience.

3.2. Elaborating the five Ws of urban resilience

Once collaborators have a common interest in, and understanding of, urban resilience, the next phase is to collectively think through questions related to resilience for whom? What? When? Where? And why? (Table 1) These "five Ws" bring the politics of resilience to the forefront by encouraging the explicit recognition of politicized decisions, scalar dimensions, and trade-offs inherent to applying resilience empirically. Who determines the resilience priorities for a city and what are their motivations for

Table 1. The five Ws of urban resilience

		Questions to Consider
Who?		Who determines what is desirable for an urban system?
		Whose resilience is prioritized?
		Who is included (and excluded) from the urban system?
What?	T	What perturbations should the urban system be resilient to?
	R	What networks and sectors are included in the urban system?
	A	Is the focus on generic or specific resilience?
When?	D	Is the focus on rapid-onset disturbances or slow-onset changes?
	E	Is the focus on short-term resilience or long-term resilience?
	O	Is the focus on the resilience of present or future generations?
Where?	F	Where are the spatial boundaries of the urban system?
	F	Is the resilience of some areas prioritized over others?
	S	Does building resilience in some areas affect resilience elsewhere?
Why?		What is the goal of building urban resilience?
		What are the underlying motivations for building urban resilience?
		Is the focus on process or outcome?

Note: Adapted from Meerow et al. (2016).

doing so? What spatial and temporal scales are included or excluded from the urban system? This section considers these and other questions related to the five Ws and the trade-offs within and between them.

When urban resilience theory is adapted to specific urban contexts, the process and outcome is highly dependent on the system and scales (e.g. temporal, spatial, jurisdictional) being included, and what disturbances or changes the system aims to become resilient to (Cutter et al., 2008; Vale, 2014; Walker & Salt, 2006). Although the resilience literature widely acknowledges that there are likely to be trade-offs in these decisions (Armitage & Johnson, 2006; Bahadur & Tanner, 2014; Fabinyi, 2008; Vale, 2014), the "nature and consequences of resilience trade-offs (between and within scales)" are still "poorly understood" (Chelleri et al., 2015, p. 182). As the remainder of this section demonstrates, considering potential trade-offs is a crucial step in thinking through each of the five Ws (Table 1).

3.2.1. Resilience for whom?

Whose vision of a desirable resilient future prevails and who benefits or loses as a result of this particular construct? Urban actors have diverse worldviews and priorities and those with the power to make decisions about how resilience is applied will do so based on their perspective. Adger (2006) and Vale (2014) suggest that decision-makers are primarily concerned with their personal short-term interests, rather than the long-term benefit of the most vulnerable. Who makes the decisions (often at a particular jurisdictional scale) thus shapes whose resilience is prioritized over what time scale (Wagenaar & Wilkinson, 2015).[4]

Who is included and excluded from the urban system of focus? Who gets to draw those boundaries? "Who counts as the city?" (Vale, 2014, p. 197). Thinking through questions of resilience for whom entails considering potential trade-offs between stakeholders (Fabinyi, 2008). As Wagenaar and Wilkinson (2015) observed in their case study of Melbourne, planning for resilience is inherently a struggle.

3.2.2. Resilience of what to what?
Operationalizing resilience requires specifying what will be made resilient to what (Carpenter et al., 2001). Urban policies and interventions vary depending on which disturbance is prioritized (e.g. climate change, natural disasters, terrorism). Enhancing resilience to military attack might require closing off access to important buildings, whereas easier entry could help aid relief efforts post disaster (Vale, 2014). Which parts of a city's population, infrastructure, or resource flows are going to be made more resilient? This entails revisiting what is included in the urban. Does it include the power plants that provide energy, for instance, if they are located outside the city proper?

A tension often exists between maximizing *specified* resilience to existing threats and *general* capacity to adapt to unanticipated disruptions (Walker & Salt, 2006, p. 121). Wu and Wu (2013) opt for general resilience based on the argument that focusing on specific threats tends to undermine the flexibility and diversity of possible system responses. Research on adaptive capacity, however, has shown that balancing the two is crucial (Eakin, Lemos, & Nelson, 2014). Chelleri and Olazabal (2012, p. 70) illustrate this potential trade-off by noting that an entirely wind-based electricity system might be a positive adaptation to current energy and climate concerns, but a more diverse and flexible energy portfolio (even including some fossil fuels) would increase the ability to adjust to future changes.

3.2.3. Resilience for when?
The wind electricity example also draws attention to temporal scale and trade-offs. Is the primary goal to build resilience to short-term disruptions (e.g. hurricanes) or long-term stress (e.g. precipitation changes caused by climate change)? If the focus is on the short term, then according to Chelleri and Olazabal (2012), the objective is system persistence, whereas a long-term perspective would likely require some degree of transition or transformation. How does building resilience for the current generation impact future ones? Walker and Salt (2006) argue that building long-term general resilience often comes at the expense of short-term efficiency. Another question related to temporal scale is whether resilience interventions focus on anticipating future threats or reacting to past disturbances (Chelleri & Olazabal, 2012; Vale, 2014).

3.2.4. Resilience for where?
Cities are inextricably linked to their surrounding regions and globally through commodity, social, economic, political, and infrastructure networks (Castells, 2002; Da Silva, Kernaghan, & Luque, 2012; Hodson & Marvin, 2010; Seitzinger et al., 2012). The resilience of a city, therefore, necessitates consideration of its relationship to larger networks of flows (Pearson & Pearson, 2014).

SES resilience theory does acknowledge the importance of cross-scalar dynamics (Bahadur & Tanner, 2014; Ernstson et al., 2010). This emphasis is represented in Gunderson and Holling (2002) influential panarchy model, where "revolt" and "remember" arrows link nested adaptive cycles (Olsson et al., 2014). These arrows indicate that local resilience may be affected by global-scale processes, such as a recession in global financial markets (Armitage & Johnson, 2006). Conversely, local-scale transformations can catalyze broader-scale change. Nevertheless, in empirical contexts, including urban applications, these scalar dimensions often receive insufficient attention (Chelleri et al., 2015; MacKinnon & Derickson, 2012). As Beilin and Wilkinson (2015, p. 4) note, where the boundary of the urban is delineated "has implications across all levels of management, government and communities." Ideally the city should be conceptualized in terms of urbanization processes that cut across scales. In practice, operationalizing resilience necessitates some limitation of spatial extent, but should at least reflect on the implications of these designations, cross-scalar interactions, and how fostering resilience at one spatial scale affects those at others.

3.2.5. Why resilience?

Given the criticism that resilience-based policies are too focused on maintaining the status quo, it becomes crucial to question why urban resilience is being studied or promoted and the ultimate goal of these interventions. Is it to improve adaptive processes generally, achieve a certain outcome, or both? Urban resilience interventions tend to prioritize swift system recovery after a disturbance, but this is not necessarily desired. As Vale (2014, p. 198) writes, "It is all too easy to talk about 'bouncing back to where we were' without asking which 'we' is counted, and without asking whether 'where we were' is a place to which a return is desirable." This connects back to the "who" questions, highlighting the need to understand the political context, decision-making processes, and powerbrokers that define the resilience agenda and to carefully consider underlying motives.

In short, urban plans and interventions must be considered in terms of political context, trade-offs, interconnections, and multiple scales. Thinking through the questions related to who, what, when, where, and why should be followed by empirical research to illuminate how these trade-offs work when resilience is operationalized in a specific context. To illustrate how differences in the five Ws shape outcomes, we briefly examine the case of green infrastructure spatial planning.

4. Urban resilience in empirical contexts

One strategy cities employ to enhance resilience is to expand green infrastructure, which Benedict and McMahon (2002, p. 12) define as: "An interconnected network of green space that conserves natural ecosystem values and functions and provides associated benefits to human populations." Based on this definition, green infrastructure includes urban green spaces such as parks, greenways, rain gardens, or green roofs (Wise, 2008). Advocates focus on the multiple social and ecological benefits of green infrastructure, from improved public health to enhanced stormwater retention (Elmqvist et al., 2015; Sussams, Sheate, & Eales, 2015; Tzoulas et al., 2007).

Green infrastructure may be particularly attractive to city officials because it provides a concrete approach for enhancing different aspects of urban resilience (Kearns, Saward, Houlston, Rayner, & Viraswamy, 2014). Depending on the technology and scale of implementation, green infrastructure can support both short- and long-term resilience through its ability to counteract the urban heat island effect, reduce the need for building cooling, reduce storm vulnerability through natural absorption of water, reduce runoff and overflows of untreated stormwater into bodies of water, and even provide a local source of food (Rouse & Bunster-Ossa, 2013). Less clear in the literature are the trade-offs between these benefits and who profits and why (Ernstson, 2013; Hansen & Pauleit, 2014; Lovell & Taylor, 2013).

Like resilience more broadly, planning for multifunctional green infrastructure requires "knowledge that crosses many disciplinary boundaries" (Kearns et al., 2014, p. 55), but getting traditionally siloed departments and agencies to work together is usually difficult (Sussams et al., 2015). Resources for urban green infrastructure (and resilience building generally) are limited, leading to difficult decisions about where to expand it. If managing stormwater is the primary determinant of where to locate new green space, for example, will it also alleviate relative park poverty? These concerns highlight the potential trade-offs between various social and environmental goals and the inherently political nature of green infrastructure planning. Thus, we briefly consider a hypothetical case of green infrastructure planning for the City of LA, which is the second largest city in the United States with a diverse population of 3.8 million living in 468 square miles (US Census, 2010). In recent years, city agencies and nongovernmental organizations have promoted green infrastructure expansion.[5] We present two hypothetical planning scenarios for LA corresponding to two desired resilience benefits or different responses to questions related to resilience for whom, what, when, where, and why (Table 2). The example shows how these choices would redraw which areas of the city are prioritized and who benefits as a result.

In hypothetical scenario #1, a municipal department (such as the Los Angeles Department of Public Works) seeks to increase resilience through better stormwater management. In scenario #2, a nongovernmental organization (such as the Trust for Public Land) aims to support community resilience by increasing access to green space. For both scenarios, existing spatial data sets are used to generate indicators for where the particular green infrastructure resilience benefit is needed most. These indicators are then aggregated and compared for each census tract within the city boundary using ArcGIS.

Table 2. Illustrative applications of the "five Ws of urban resilience" to green infrastructure planning

		Scenario 1	Scenario 2
Who?		Beneficiaries are city residents living in flood risk zones	Beneficiaries are city residents with most limited access to green space
What?	T R	Specifically focused on stormwater management	Generic community resilience
When?	A D	Focused on current residents and based on current estimates of risk	Both short-term and long-term resilience
Where?	E O F F	Neighborhoods with the most area in flood hazard zones within the municipal boundaries	Neighborhoods with the lowest average access to green space (parks) within the municipal boundaries
Why?	S	Goal is an outcome: Flood losses and investments in 'grey' stormwater infrastructure are reduced	Goal is an outcome: increased social justice

4.1. *Hypothetical scenario #1: optimizing green infrastructure for stormwater management*

The first scenario focuses on the stormwater management benefits of green infrastructure, historically the predominant rationale for its deployment (Newell et al., 2013). The goal is to build resilience through improved stormwater management, and in this case, flood risk maps are used as a spatial indicator for where stormwater is likely to accumulate. Consequently, the chief beneficiaries are residents living in these areas. Priority areas for stormwater management are based on 2008 Federal Emergency Management Agency National Flood Hazard Layer Flood Insurance Rate Maps for Los Angeles County. High-risk areas (1% annual chance of flood hazard) and medium-risk areas (0.2% annual chance of flood hazard) are merged. The final tract score is a function of the area of this flood hazard layer within (intersecting) the tract.

4.2. *Hypothetical scenario #2: optimizing green infrastructure to increase access to green space*

Access to green space is associated with many social benefits and increased community resilience, which is why cities like LA may aim to increase social equity with respect to green space access (Tidball & Krasny, 2014; Wolch, Byrne, & Newell, 2014). In this scenario, green infrastructure development is prioritized for neighborhoods that have relative park poverty as a proxy for access to green space. This scenario thus promotes generic community resilience through more equitable green space distribution. To identify areas of park poverty, we use a GIS data set containing all the parks in Los Angeles that was generated as part of the 2008 Green Visions Plan (Newell et al., 2007). A quarter-mile buffer is drawn around each park, and this area denoted as accessible park acreage (Wolch, Wilson, & Fehrenbach, 2005). To determine the average amount of accessible park area per person for each census tract, the total accessible park area intersecting each tract is divided by the population living in that tract. The resulting attribute is the basis of the park poverty indicator.

4.3. *Comparing green infrastructure scenarios*

Reflecting on the five Ws (Table 2), the two scenarios generate very different spatial outcomes, providing different benefits to communities in these areas. The motivation (or why question) for the green infrastructure differs, reflecting the interests of the actors setting the agenda. In scenario #1, individuals located in areas of high flood risk are likely to benefit. Scenario #2 would focus green infrastructure development in neighborhoods with smaller park acreage per resident in an effort to address inequalities in access to green space. Concerns related to spatial scale come up in both scenarios. In scenario #1 and #2, the system boundary is the City of LA, thus the residents living within its boundaries would benefit more directly, rather than the larger metropolitan area. Census tracts are the basic unit of analysis, but with an average population of 4000 they are likely heterogeneous, and this variation may not be accurately represented by tract-level data. For example, if there is a large park on one side of a tract, the park poverty score may be low, even if residents on the other side of the tract have no accessible park area.

The scenarios also differ in terms of what is being made resilient to what. The first is aimed at building resilience to a specific challenge (e.g. stormwater management), whereas the second seeks to foster generic community resilience through more equitable distribution of green space. With regard to temporal scale, both are similarly focused on current populations rather to past or future generations. For example, scenario #1 uses current estimates of flood risk rather than future risk profiles based on long-term climate impacts.

Figure 4 illustrates how different areas of LA would be prioritized for green infrastructure in the two hypothetical scenarios. In both cases, standardized census tract indicator values are divided into 10 quantiles, with a score of 1 representing "low priority" and 10 "high priority." The statistically significant negative correlation[6] between the tract values in the two scenarios indicates that spatial trade-offs are involved. If flood risk is the primary determinant, then it may not address other resilience needs. If green infrastructure is only developed in flood hazard zones in LA, environmental justice advocates concerned with park poverty might be less willing to provide support than if it were implemented in their priority areas. One possible solution might be to layer different criteria and identify spatial "hotspots" (i.e. areas where green infrastructure benefits can be coupled). A wide range of stakeholders could then be asked to weight the importance of the criteria for siting it, and these weights used to develop combined planning scenarios.

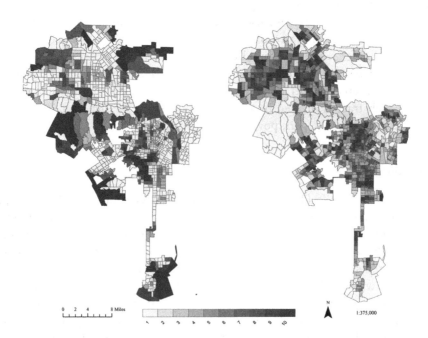

Figure 4. Priority census tracts for green infrastructure development in the City of Los Angeles based stormwater management (left) and access to green space (right)
Note: Maps show standardized census tract scores divided into 10 quantiles. Darker colors indicate higher priority.

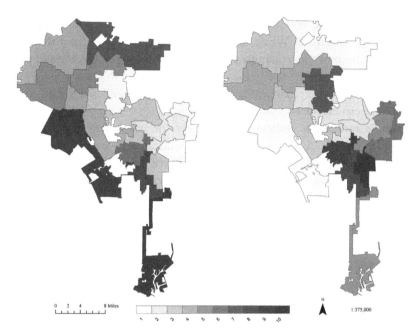

Figure 5. Priority council districts for green infrastructure development in the City of Los Angeles based on stormwater management (left) and access to green space (right)
Note: Standardized tract scores are aggregated at the council district level, and these district scores divided into 10 quantiles. Darker colors indicate higher priority.

The scale of analysis (and scale at which planning decisions are made) has implications for what gets prioritized and where. When the scenario scores are aggregated to the LA Council District scale (Figure 5), different trade-off patterns emerge. While a negative relationship still persists between stormwater management and park access, it is no longer statistically significant. When comparing the results of the two scenarios at the scale of the census tract and council district, priority hotspots that appear in the census tract analysis are obscured in the council district analysis.

This brief example provides a basic illustration of how spatial planning based on different resilience benefits, and at different scales, would impact priorities for green infrastructure development. It, therefore, highlights the challenges associated with planning for urban resilience, the likelihood of inherent trade-offs in this process, and the need to critically examine the politics and practices of resilience planning to determine whose priorities are being implemented and at what cost. Every resilience planning or measurement decision is inevitably a political one, with winners and losers, thus resilience needs be operationalized through a collaborative and inclusive process that takes into account varying stakeholder priorities.

5. Conclusion

Resilience theory has evolved into an influential global discourse, including for urban research and policy. For some, resilience is eclipsing sustainability, vulnerability, and adaptation as the primary organizing principle for managing the unpredictable and changing futures of SESs, including cities(Davoudi et al., 2012; Elmqvist, Barnett, & Wilksonson, 2014; O'Hare & White, 2013). As the popularity of the urban resilience concept grows, it becomes increasingly important to interrogate the ways in which it is used. Social scientists have made significant contributions to this discourse by critically evaluating the term's conceptual ambiguities, conservative tendencies, and underdeveloped usage in social contexts.

The paper introduces a collaborative process for advancing a *politics of urban resilience*, which entails confronting inherent political and scalar complexities and trade-offs. We have divided this process into three phases: urban resilience as a boundary object, the five Ws of urban resilience, and urban resilience in empirical contexts. To highlight trade-offs and policy implications related to the five Ws and the politics of urban resilience, we provided two potential scenarios of green infrastructure spatial planning in Los Angeles. This brief example illustrated how prioritizing one resilience benefit of green infrastructure (e.g. stormwater abatement) over another (e.g. alleviating park poverty) could lead to markedly different spatial priorities, with implications for a city's ecology and socioeconomic fabric. This suggests a need for future research to scrutinize resilience-building planning decisions and the ways in which different models of decision-making affect outcomes.

Critical human geographers were among the first scholars to interrogate the growing influence of resilience discourse, contributing to a richer understanding of the concept's limitations. This provides a foundation for additional investigations into, for example, issues of power and how disparities might impact even the most collaborative resilience decision-making, which has been understudied in the resilience literature (Olsson et al., 2014). Urban political ecologists could contribute by continuing to ask "questions about who produces what kind of social-ecological configurations for whom" (Heynen, Kaika, & Swyngedouw, 2006, p. 2). The urban resilience literature needs a more nuanced appreciation for what defines the "city" or "urban," as well as attentiveness to scalar dimensions. Finally, geographers can continue to provide empirically rich place-based research that advances our understanding of what resilience means and how it is applied in different urban contexts.

Notes

1. Examples of international resilience policy initiatives include the Rockefeller Foundation's "100 Resilient Cities" campaign, the United Nations Office for Disaster Risk Reduction's "Making Cities Resilient" program, and ICLEI's "Resilient Cities" program.
2. Why resilience seems to have become more of a buzzword than vulnerability or adaptation is unclear. One explanation is that resilience is more politically tractable than vulnerability or adaptation simply because of its positive connotation (McEvoy, Fünfgeld, & Bosomworth, 2013; O'Hare & White, 2013; Sudmeier-Rieux, 2014).
3. Vale (2014) provides a helpful anecdote: In Sri Lanka, poor fishing villages were relocated inland following the 2004 tsunami, and more robust hotel structures built in their place. If the "system" is defined as the entire city, this would seem a positive development, but closer examination reveals that wealthy hotel owners benefitted, while the fishing communities lost their livelihoods.
4. The question of resilience for whom has obvious relevance to nonhuman actors. As Beilin and Wilkinson (2015, p. 3) write, "We cannot ignore the non-human species encapsulated

within the territory of and significantly affected by the ever-expanding urban or its amorphous boundaries."
5. The city has a number of plans and initiatives including the Green Streets program and the Emerald Necklace Forest to Ocean Extended Vision Plan (Goodyear, 2014).
6. Pearson's correlation coefficient is −0.07, which is significant at $p < 0.05$.

Acknowledgements

The authors would like to thank the editors of this special issue, V. Kelly Turner and David Kaplan, the anonymous reviewers, and the members of the University of Michigan Interdisciplinary Workshop on Urban Sustainability and Resilience for feedback on an early draft.

Disclosure statement

No potential conflict of interest was reported by the authors.

ORCID

Sara Meerow ⓘ http://orcid.org/0000-0002-6935-1832
Joshua P. Newell ⓘ http://orcid.org/0000-0002-1440-8715

References

Adger, W. Neil. (2006). Vulnerability. *Global Environmental Change, 16*(3), 268–281.
Armitage, Derek, & Johnson, Derek. (2006). Can resilience be reconciled with globalization and the increasingly complex conditions of resource degradation in Asian Coastal Regions? *Ecology and Society, 11*(1), 1–19.
ARUP. (2014). *City resilience index: City resilience framework*. London: Ove Arup & Partners International Limited.
Bahadur, Aditya, & Tanner, Thomas (2014). Transformational resilience thinking: Putting people, power and politics at the heart of urban climate resilience. *Environment and Urbanization, 26*(1), 200–214.
Batty, Michael. (2008). The size, scale, and shape of cities. *Science, 319*(5864), 769–771.
Beichler, Simone, Hasibovic, Sanin, Davidse, Bart Jan, & Deppisch, Sonja. (2014). The role played by social-ecological resilience as a method of integration in interdisciplinary research. *Ecology and Society, 19*(3), 1–8.
Beilin, Ruth, & Wilkinson, Cathy. (2015). Introduction: Governing for urban resilience. *Urban Studies, 52*(7), 1205–1217.
Benedict, Mark A., & McMahon, Edward T. (2002). Green infrastructure: Smart conservation for the 21st century. *Renewable Resources Journal, 20*(3), 12–17.
Brand, Fridolin Simon, & Jax, Kurt. (2007). Focusing the meaning(s) of resilience: Resilience as a descriptive concept and a boundary object. *Ecology and Society, 12*(1), 1–23.
Brenner, Neil, & Schmid, Christtian. (2011). Planetary urbanization. In Matthew Gandy (Ed.), *Urban constellations* (pp. 10–14). Berlin: Jovis.
Brown, Katrina. (2012). Policy discourses of resilience. In M. Pelling, D. Manuel-Navarrete, & M. Redclift (Eds.), *Climate change and the crisis of capitalism: A chance to reclaim self, society and nature* (p. 37). Oxon: Routledge.
Brown, Katrina. (2014). Global environmental change I: A social turn for resilience? *Progress in Human Geography, 38*(1), 107–117.

Bruntland, Gro Harlem. (1987). *Our common future our common future: Report of the 1987 World Commission on Environment and Development*. Oxford: Oxford University Press.
Carpenter, Steve, Walker, Brian, Anderies, J. Marty, & Abel, Nick. (2001). From metaphor to measurement: Resilience of what to what ? *Ecosystems*, 4(8), 765–781.
Cascio, Jamais. (2009). The next big thing: Resilience. *Foreign Policy*, 172, 92.
Castells, Manuel. (2002). The space of flows. In Ida Susser (Ed.), *The castells reader on cities and social theory* (p. 448). Malden, MA: Blackwell.
Chelleri, Lorenzo, & Olazabal, Marta. (2012). *Multidisciplinary perspectives on urban resilience*. Bibao, Spain: Basque Centre for Climate Change.
Chelleri, Lorenzo, Waters, J. J., Olazabal, Marta, & Minucci, Guido. (2015). Resilience trade-offs: Addressing multiple scales and temporal aspects of urban resilience. *Environment and Urbanization*, 27(1), 181–198.
Coaffee, Jon. (2008). Risk, resilience, and environmentally sustainable cities. *Energy Policy*, 36 (12), 4633–4638.
Coaffee, Jon. (2013). Towards next-generation urban resilience in planning practice: From securitization to integrated place making. *Planning Practice and Research*, 28(3), 323–339.
Cote, Muriel, & Nightingale, Aandrea J. (2011). Resilience thinking meets social theory: Situating social change in socio-ecological systems (SES) research. *Progress in Human Geography*, 36(4), 475–489.
Cretney, Raven. (2014). Resilience for whom? Emerging critical geographies of socio-ecological resilience. *Geography Compass*, 8(9), 627–640.
Cutter, Susan, Barnes, Lindsey, Berry, Melissa, Burton, Christopher, Evans, Elijah, Tate, Eric, & Webb, Jennifer. (2008). A place-based model for understanding community resilience to natural disasters. *Global Environmental Change*, 18(4), 598–606.
Da Silva, Jo, Kernaghan, Sam, & Luque, Andrés. (2012). A systems approach to meeting the challenges of urban climate change. *International Journal of Urban Sustainable Development*, 4 (2), 125–145.
Davoudi, Simin, Shaw, Keith, Haider, L. Jamila, Quinlan, Allyson E., Peterson, Garry D., Wilkinson, Cathy, & Porter, Libby. (2012). Resilience: A bridging concept or a dead end? "Reframing" resilience: Challenges for planning theory and practice interacting traps: Resilience assessment of a pasture management system in Northern Afghanistan urban resilience: What does it mean in planning practice? Resilience as a useful concept for climate change adaptation? The politics of resilience for planning: A cautionary note. *Planning Theory & Practice*, 13(2), 299–333.
Derissen, Sandra, Quaas, Martin, & Baumgärtner, Stefan. (2011). The relationship between resilience and sustainability of ecological-economic systems. *Ecological Economics*, 70(6), 1121–1128.
Desouza, Kevin, & Flanery, Trevor H (2013). Designing, planning, and managing resilient cities: A conceptual framework. *Cities*, 35, 89–99.
Dicken, Peter (2011). *Global shift: Mapping the changing contours of the world economy*. New York: Guilford Press.
Dilling, Lisa, & Lemos, Maria Carmen (2011). Creating usable science: Opportunities and constraints for climate knowledge use and their implications for science policy. *Global Environmental Change*, 21(2), 680–689.
Eakin, Hallie C., Lemos, Maria C., & Nelson, Donald R (2014). Differentiating capacities as a means to sustainable climate change adaptation. *Global Environmental Change*, 27, 1–8.
Elmqvist, Thomas. (2014). Urban resilience thinking. *Solutions*, 5(5), 26–30.
Elmqvist, Thomas, Barnett, Guy, & Wilkinson, Cathy. (2014). Exploring urban sustainability and resilience. In Leonie J. Pearson, Peter W. Newman, & Peter Roberts (Eds.), *Resilient sustainable cities: A future* (pp. 19–28). New York, NY: Routledge.
Elmqvist, Thomas, Setälä, H., Handel, S. N., Van Der Ploeg, S., Aronson, J., Blignaut, J. N., & De Groot, R. (2015). Benefits of restoring ecosystem services in urban areas. *Current Opinion in Environmental Sustainability*, 14, 101–108.

Ernstson, Henrik. (2013). The social production of ecosystem services: A framework for studying environmental justice and ecological complexity in urbanized landscapes. *Landscape and Urban Planning, 109*(1), 7–17.

Ernstson, Henrik, Van Der Leeuw, Sander, Redman, Charles, Meffert, Douglas, Davis, George, Alfsen, Christine, & Elmqvist, Thomas. (2010). Urban transitions: On urban resilience and human-dominated ecosystems. *Ambio, 39*(8), 531–545.

Evans, Brad, & Reed, Julian. (2014). *Resilient life: The art of living dangerously*. Cambridge, UK: Polity Press.

Evans, James P. (2011). Resilience, ecology and adaptation in the experimental city. *Transactions of the Institute of British Geographers, 36*, 223–237.

Fabinyi, Michael. (2008). *The political aspects of resilience*. Proceedings of the 11th International Coral Reef Symposium, Fort Lauderdale, FL, pp. 971–975.

Folke, Carl. (2006). Resilience: The emergence of a perspective for social–ecological systems analyses. *Global Environmental Change, 16*(3), 253–267.

Folke, Carl, Carpenter, Steve, Elmqvist, Thomas, Gunderson, Lance, Holling, C. S., & Walker, Brian. (2002). Resilience and sustainable development: Building adaptive capacity in a world of transformations. *AMBIO: A Journal of the Human Environment, 31*(5), 437–440.

Friend, Richard, & Moench, Marcus. (2013). What is the purpose of urban climate resilience? Implications for addressing poverty and vulnerability. *Urban Climate, 6*, 98–113.

Gallopín, Gilberto C. (2006). Linkages between vulnerability, resilience, and adaptive capacity. *Global Environmental Change, 16*(3), 293–303.

Garschagen, Matthias. (2013). Resilience and organisational institutionalism from a cross-cultural perspective: An exploration based on urban climate change adaptation in Vietnam. *Natural Hazards, 67*(1), 25–46.

Goodyear, Sarah. (2014, August). Ambitious L.A. Parks plan will require coordination of 88 cities. *Next City*. Retrieved from https://nextcity.org/daily/entry/los-angeles-emerald-necklace-plan-la-basin

Gunderson, Lance, & Holling, Crawford Stanley. (2002). *Panarchy: Understanding transformations in human and natural systems*. (L. Gunderson & C. S. Holling, Eds.). Washington, DC: Island Press.

Hansen, Rieke, & Pauleit, Stephan. (2014). From multifunctionality to multiple ecosystem services? A conceptual framework for multifunctionality in green infrastructure planning for urban areas. *Ambio, 43*(4), 516–529.

Harvey, David. (1996). *Justice, nature, and the geography of difference*. London, UK: Blackwell.

Heynen, Nik, Kaika, Maria, & Swyngedouw, Erik. (2006a). Urban political ecology. In Nik Heynan, Maria Kaika, & Erik Swyngedouw (Eds.), *In the nature of cities: urban political ecology and the politics of urban metabolism*. New York: Routledge.

Hodson, Mike, & Marvin, Simon. (2010). Can cities shape socio-technical transitions and how would we know if they were? *Research Policy, 39*(4), 477–485.

Holling, Crawford Stanley. (1973). Resilience and stability of ecological systems. *Annual Review of Ecology and Systematics, 4*, 1–23.

Holling, Crawford Stanley. (1996). Engineering resilience versus ecological resilience. In Peter Schulze (Ed.), *Engineering withn ecological constraints*. Washington, DC: The National Academies Press.

Jabareen, Yosef. (2013). Planning the resilient city: Concepts and strategies for coping with climate change and environmental risk. *Cities, 31*, 220–229.

Janssen, Marcoa, Schoon, Michael, Ke, Weimao, & Börner, Katy. (2006). Scholarly networks on resilience, vulnerability and adaptation within the human dimensions of global environmental change. *Global Environmental Change, 16*(3), 240–252.

Jasanoff, Sheila. (2004). *States of knowledge: The co-production of science and social order*. New York, NY: Routledge.

Joseph, Jonathan. (2013). Resilience as embedded neoliberalism: A governmentality approach. *Resilience: International Policies, Practices and Discourses, 1*(1), 38–52.

Kearns, Allen, Saward, Rhiannon, Houlston, Alex, Rayner, John, & Viraswamy, Harry. (2014). Building urban resilience through green infrastructure pathways. In Leonie J. Pearson, Peter W. Newman, & Peter Roberts (Eds.), *Resilient sustainable cities: A future* (pp. 52–65). New York, NY: Routledge.

Korhonen, Jouni, & Seager, Thomas P. (2008). Beyond eco-efficiency: A resilience perspective. *Business Strategy and the Environment, 17*, 411–419.

Leach, Melissa. (2008). *Re-framing resilience: A symposium report* (Steps Working Paper No. 13). Brighton. Retrieved from http://steps-centre.org/wp-content/uploads/Resilience.pdf

Lebel, Louis, Anderies, John M., Campbell, Bruce, Folke, Carl, Hatfield-dodds, Steve, Hughes, Terry P., & Wilson, James. (2006). Governance and the capacity to manage resilience in regional social-ecological systems. *Ecology and Society, 11*(1), 1–21.

Leichenko, Robin. (2011). Climate change and urban resilience. *Current Opinion in Environmental Sustainability, 3*(3), 164–168.

Lovell, Sarah, & Taylor, John R. (2013). Supplying urban ecosystem services through multi-functional green infrastructure in the United States. *Landscape Ecology, 28*, 1447–1463.

MacKinnon, D., & Derickson, K. D. (2012). From resilience to resourcefulness: A critique of resilience policy and activism. *Progress in Human Geography, 37*(2), 253–270.

Matyas, David, & Pelling, Mark. (2014). Positioning resilience for 2015: The role of resistance, incremental adjustment and transformation in disaster risk management policy. *Disasters, 39*(S1), S1–S18.

McEvoy, Darryn, Fünfgeld, Hartmut, & Bosomworth, Karyn. (2013). Resilience and climate change adaptation: The importance of framing. *Planning Practice and Research, 28*(3), 280–293.

Meerow, Sara, & Baud, Isa. (2012). Generating Resilience: Exploring the contribution of the small power producer and very small power producer programs to the resilience of Thailand's power sector. *International Journal of Urban Sustainable Development, 4*(1), 20–38.

Meerow, Sara, & Newell, Joshua P. (2015). Resilience and complexity: A bibliometric review andprospects for industrial ecology. *Journal of Industrial Ecology, 19*(2), 236–251.

Meerow, Sara, Newell, Joshua P., & Stults, Melissa. (2016). Defining urban resilience: A review. *Landscape and Urban Planning, 147*, 38–49.

Miller, Fiona, Osbahr, Henny, Boyd, Emily, Thomalla, Frank, Bharwani, Sukaina, Ziervogel, Gina, & Nelson, Donald. (2010). Resilience and vulnerability: Complementary or conflicting concepts? *Ecology and Society, 15*(3), 1–11.

Newell, Joshua P., Sister, Chona, Wolch, Jennifer, Swift, Jennifer, Ghaemi, Parisa, Wilson, John, & Longcore, Travis. (2007). *Creating parks & open space using green visions planning toolkit 1.0*. Los Angeles: University of Southern California GIS Research Laboratory and Center for Sustainable Cities.

Newell, Joshua P., Seymour, Mona, Yee, Thomas, Renteria, Jennifer, Longcore, Travis, Wolch, Jennifer, & Shishkovsky, Anne. (2013). Green alley programs: Planning for a sustainable urban infrastructure? *Cities, 31*, 144–155.

O'Hare, Paul, & White, Iain. (2013). Deconstructing resilience: Lessons from planning practice. *Planning Practice and Research, 28*(3), 275–279.

Olsson, Per, Galaz, Victor, & Boonstra, Wiebren J. (2014). Sustainability transformations: A resilience perspective. *Ecology and Society, 19*(4), 1–13.

Pearson, Leonie J, & Pearson, Craig. (2014). Adaptation and transformation for resilient and sustainable cities. In Leonie J. Pearson, Peter W. Newman, & Peter Roberts (Eds.), *Resilient sustainable cities: A future* (pp. 242–248). New York, NY: Routledge.

Pickett, STA, Cadenasso, ML, Grove, M, Nilon, C, Pouyat, R, Zipperer, WC., & Costanza, Robert (2001). Urban ecological systems: Linking terrestrial ecological, physical, and socio-economic components of metropolitan areas. *Annual Review of Ecology and Systematics, 32*(1), 127–157.

Pizzo, Barbara. (2015). Problematizing resilience: Implications for planning theory and practice. *Cities, 43*, 133–140.

Redman, Charles L. (2014). Should sustainability and resilience be combined or remain distinct pursuits? *Ecology and Society, 19*(2), 1–37.

Rouse, David C, & Bunster-Ossa, Ignacio F. (2013). *Green infrastructure: A landscape approach.* Chicago: American Planning Association.

Seitzinger, Sybil P, Svedin, Uno, Crumley, Carole L, Steffen, Will, Abdullah, Saiful Arif, Alfsen, Christine, & Sugar, Lorraine. (2012). Planetary stewardship in an urbanizing world: Beyond city limits. *Ambio, 41*(8), 787–794.

Seto, Karen, Sánchez-Rodríguez, Roberto, & Fragkias, Michail. (2010). The new geography of contemporary urbanization and the environment. *Annual Review of Environment and Resources, 35*(1), 167–194.

Star, Susan L., & Griesemer, James R. (1989). Institutional ecology, "Translations" and Boundary Objects: Amateurs and Professionals in Berkeley's Museum of Vertebrate Zoology, 1907-39. *Social Studies of Science, 19*(3), 387–420.

Sudmeier-Rieux, Karen I. (2014). Resilience – an emerging paradigm of danger or of hope? *Disaster Prevention and Management, 23*(1), 67–80.

Sussams, LW, Sheate, WR, & Eales, RP. (2015). Green infrastructure as a climate change adaptation policy intervention: Muddying the waters or clearing a path to a more secure future? *Journal of Environmental Management, 147,* 184–193.

Tidball, Keith G, & Krasny, Marianne E. (2014). Introduction: Greening in the Red Zone. In Keith G. Tidball & Marianne E. Krasny (Eds.), *Greening in the Red Zone - Disaster, resilience and community greening.* Dordrecht, Netherlands: Springer Netherlands.

Turner, Billy Lee. (2010). Vulnerability and resilience: Coalescing or paralleling approaches for sustainability science? *Global Environmental Change, 20*(4), 570–576.

Turner, Matthew D. (2014). Political ecology I: An alliance with resilience? *Progress in Human Geography, 38*(4), 616–623.

Tzoulas, Konstantinos, Korpela, Kalevi, Venn, Stephen, Yli-Pelkonen, Vesa, Kaźmierczak, Aleksandra, Niemela, Jari, & James, Philip. (2007). Promoting ecosystem and human health in urban areas using Green Infrastructure: A literature review. *Landscape and Urban Planning, 81*(3), 167–178.

United States Census. (2010). Retrieved from www.census.gov.

Vale, Lawrence J. (2014). The politics of resilient cities: Whose resilience and whose city? *Building Research & Information, 42*(2), 191–201.

Wagenaar, Hendrik, & Wilkinson, Cathy. (2015). Enacting resilience: A performative account of governing for urban resilience. *Urban Studies, 52*(7), 1265–1284.

Walker, Brian, & Salt, David. (2006). *Resilience thinking: Sustaining ecosystems and people in a changing world.* Washington, DC: Island Press.

Walker, Jeremy, & Cooper, Melinda. (2011). Genealogies of resilience: From systems ecology to the political economy of crisis adaptation. *Security Dialogue, 42*(2), 143–160.

Wamsler, Christine. (2014). *Cities. Disaster risk and adaptation.* New York, NY: Routledge.

Weichselgartner, Jurgen, & Kelman, Ilan. (2015). Geographies of resilience: Challenges and opportunities of a descriptive concept. *Progress in Human Geography, 39*(3), 249–267.

Welsh, Marc. (2014). Resilience and responsibility: Governing uncertainty in a complex world. *The Geographical Journal, 180*(1), 15–26.

Wise, Steve. (2008). Green infrastructure rising. *Planning, 74*(8), 14–19.

Wolch, Jennifer, Wilson, John, & Fehrenbach, Jed. (2005). Parks and park funding in Los Angeles: An equity-mapping analysis. *Urban Geography, 26*(1), 4–35.

Wolch, Jennifer, Byrne, Jason, & Newell, Joshua P. (2014). Urban green space, public health, and environmental justice: The challenge of making cities "just green enough.". *Landscape and Urban Planning, 125,* 234–244.

Wu, Jianguo, & Wu, Tong. (2013). Ecological resilience as a foundation for urban design and sustainability. In T. A. Steward, M. L. Cadenasso Pickett, & Brian McGrath (Eds.), *Resilience in ecology and urban design: Linking theory and practice for sustainable cities* (pp. 211–230). Dordrecht, Netherlands: Springer.

Green infrastructure, green space, and sustainable urbanism: geography's important role

Lisa Benton-Short, Melissa Keeley and Jennifer Rowland

ABSTRACT
This paper is a broad review of green infrastructure theory and practice relative to urban sustainability and the space for geographers in these discussions. We use examples from various urban sustainability plans to highlight ways in which green infrastructure is being conceptualized and implemented. We explore how geography contributes research on green infrastructure as well as the emerging practices as seen within sustainability plans. We identify four areas in which geographers can influence both green infrastructure theory and practice: 1) scale; 2) mapping distribution; 3) sensitivity to place and locale; and 4) equity and access. We conclude that in these areas geographers have tremendous opportunity contribute more deliberately to sustainable urbanism.

Introduction

In recent years and for a variety of reasons, cities have taken the lead in sustainability efforts in the United States (Portney, 2003; Tang et al., 2010). Urban geographers have long recognized that the city scale can be a beneficial starting point for local activism and community involvement around sustainability (Boone & Moddares, 2006; Gandy, 2002; Robbins, 2012; Short, 1989). Indeed, cities have tremendous control over land use, public education and economic development, meaning local policy can have a significant impact (Wheeler, 2013). As federal leadership in the creation of environmental policies has faltered, there has been growing support for local initiatives, often referred to as a "new localism" in environmental policies (Parker & Rowlands, 2007; Portney, 2003).

Since 2000, many cities around the world have developed comprehensive sustainability plans. Municipal sustainability plans are comprehensive visions and goals set forth by a government or other civic organization. A municipal sustainability plan is a holistic and multi-departmental document that outlines a city's goals, visions and priorities for a sustainable future. Such plans cover a diverse range of issues: climate, energy, transportation, green jobs, housing, human health, recreation and parks, etc. These plans often inventory current problems and conditions, identify solutions and priorities and set indicators for measuring progress (Evenson, Aytur, Daniel AS., & Salvesen, 2009). Today, more than 50 U.S. cities have sustainability plans, including large cities like New York City and Chicago, medium-sized cities like Cincinnati, Chattanooga, Portland, and Salt Lake City and small cities like Topeka and Burlington.

However, because there is no national standard for sustainability plans in the U.S. and because priorities can be so varied across different cities and geographies, the plans tend to differ in the way they are created and organized, the topic areas they contain or omit, and the regional and local-specific problems they aim to solve. Some plans focus on the issue of climate change, while others may focus more on local environmental, economic or equity issues or broad community goals (Portney, 2003; Saha, 2009).[1] Of particular interest is that some plan elements, including green infrastructure and green space, often reach across many elements of a city's sustainability plan. Green space is often discussed in plan sub-sections as diverse as climate adaptation, health and recreation and food resources. Both the cross-cutting nature of green infrastructure and the novelty of these planning practices make an examination of green infrastructure within urban sustainability plans a particularly insightful way to understand how this emerging and evolving approach is being utilized by cities.

This paper is a broad review of the state of green infrastructure theory and practice within the context of urban sustainability. We use municipal sustainability plans to discuss how cities are undertaking green space planning and implementation. Both the newness of sustainability plans and their lack of standardization make these documents telling indicators of how cities are conceptualizing and prioritizing green infrastructure implementation.

We focus here on small-scale green infrastructure types generally within urban fabric including parks, tree canopy, and community gardens. As opposed to large-scale infrastructure such as wetlands reconstruction or green parkway systems, we focus on small-scale green infrastructure because it is integrated within the urban fabric and therefore has the potential to impact multiple areas of a city. Small-scale infrastructure also interacts with other forms of infrastructure and can engage residents and business owners in its implementation and maintenance.

We begin with a brief overview of green infrastructure. We then continue our review of green infrastructure and green space theory and practice in the context of four key geographic concepts: 1) scale, 2) mapping distribution, 3) place and locale and 4) equity and access. We conclude that geographers, with spatial and scalar skills in landscape analysis, are uniquely positioned to contribute to advancing urban sustainability planning in these important ways.

Part I. Green infrastructure

"Green infrastructure" is a term that has come into common use within the last twenty years. Generally, this this term has gained prominence as a way to talk about the services provided by natural systems and to convey the need to manage and maintain these systems like other types of infrastructure – roads, electrical lines, water distribution systems.

However, green infrastructure is a highly contested term and has multiple definitions. Some scholars, such as Bolund and Hunhammar (1999) broadly define green infrastructure as vegetation, soils, and bioengineered systems that provide ecological services such as microclimate regulation, air quality improvements, habitat, and stormwater management. Benedict and McMahon (2006) updated this definition along similar lines calling green infrastructure ""an interconnected network of natural areas and other open spaces that conserves natural ecosystem values and functions, sustains clean air and water, and provides a wide array of benefits to people and wildlife"' (Benedict and McMahon, 2006, p. 12). However, others utilize the term much more specifically, as a synonym for stormwater Best Management Practices, referring to green roofs, permeable pavements, rain gardens, stormwater treatment swales and the like (EPA, 2014). Along these lines, Mell (2012) notes that there are substantial regional variation in the focus of green infrastructure programs.

There is also great variety in the scale of green infrastructure. Weber, Sloan, and Wolf (2006) and others discuss green infrastructure as interconnected networks of large parks, greenbelts, and wetlands. Others focus on small-scale green amenities that can improve the urban environment and provide aesthetic amenities when "decentralized" and distributed throughout the cityscape (Tzoulas et al., 2007). Examples include street trees and other vegetation, green roofs, green façades, permeable pavements. Such decentralized efforts provide a "greening" option even in dense and largely built-out urban areas where available land is scarce or expensive (Montalto et al., 2007). Indeed, green infrastructure should "operate at all spatial scales from urban centres to the surrounding countryside" and offers differentiated functionality dependent on location (Gill, Handley, Ennos, & Pauleit, 2007, p. 116).

There remains ambiguity in the term green infrastructure generally and its component practices (e.g green roofs and bioretention) (Young, Zanders, Lieberknecht, & Fassman-Beck, 2014). Ignatieva, Stewart, and Meurk (2011) noted the variety of terms used to describe green areas in cities within the fields of urban planning and design. These include green infrastructure, but commonly include terms such as "'urban open space'", "'urban green space'" and "'public open space.'" Standardization of these terms is challenging given regional variation, yet Young et al. (2014) attempted a standardization of green infrastructure typologies to promote improved benchmarking and lesson learning.

It is important to note that research in green infrastructure and green space is occurring in several disciplines. Three disciplines – geography, ecology and urban planning – form much of the research around the urban-environmental nexus, and hence green infrastructure and green space.

Urban geographers and urban planners have long shared mutual interests around cities, land use and planning. Urban planners have contributed to research on the design and use of green space and green infrastructure. Their work has also contributed to examinations of the economic benefits of green space and their role in regional development (see for example, Foster, Lowe, & Winkelman, 2011; Horwood, 2011; Netusil, Levin, Shandas, & Hart, 2014; Schäffler & Swilling, 2013; Schilling & Logan, 2008; Vandermueluen, Verspecht, Vermeire, Van Huylenbroeck, & Gellynck, 2011).

Ecologists, who have largely preferred to exclude humans from their study of natural systems, are now turning their attention to the urban-environmental nexus examining concepts of energetics, material budgets, and the metabolism of the city as a whole (Coleman, 2010; Collins et al., 2000; Palmer & Ruhl, 2015; Pataki et al., 2011; Pickett & Cadenasso, 2002). Research has emphasized the environmental services of green infrastructure, such as water retention or biodiversity. Such work is exemplified by the two NSF urban Long Term Ecological Research (LTER) programs in Baltimore, Maryland and Phoenix, Arizona.[2] These sites have given us some insight into the ecosystem services and benefits of green infrastructure.

The study of urban environments has evolved, initially from a focus on wildlife inventories and botanical surveys. Today, the emphasis in urban ecology is "the study of spatiotemporal environmental patterns and impacts." (McDonnell, 2015). Research is focusing more complex urban systems such as streams and stormwater drainage networks, to vacant lots and lawns and how these systems interact and how they create feedbacks and exchanges. Current research includes broader spatial scales, many patch types and large distances. As the field of urban ecology has evolved, the focus has shifted to spatial complexity and more emphasis on seeing urban systems as internally complex and dynamic. More recently geographers have been studying interactions between social and biogeophysical structures to better understand the feedbacks and exchanges among systems. Examining land cover change reflects this new complexity (see Zhou, Cadenasso, Schwarz, & Pickett, 2014). In urban ecology, geographers have helped to advance both spatial theory and spatial tools.

McDonnell notes that there remains a huge demand for spatially informed research about urban ecosystems – including urban climate, soils and vegetation, green spaces, planning and management (2015). Explaining the distribution and abundance in organisms is a critical area where there are still knowledge gaps. For example, Gagne and colleagues look at scale mismatches between science and practice in urban landscapes (Gagné et al., 2015). In addition, even linear infrastructure or small patches such as roads, railways or small patches such as vacant land can provide ecological services and value. Geospatial analysis can provide a framework for understanding these spatial patterns within multiple areas within a city. Seto, Güneralp, and Hutrya (2012) examine how urban expansion and development may impact urban land-cover change (and hence carbon storage) by developing spatially explicit probabilistic forecasts of global urban land-cover change . McPhearson, Kremer, and Hamstead (2013) mapped ecosystem services in New York City to better understand how local and regional trends and plans affect ecosystem provisioning. Zhou, Huang, Pickett, and Cadenasso (2011) used historic maps and aerial photographs to compare patterns of forest cover in metropolitan Baltimore from 1914–2004. They found that while the total forest area was stable, forest patches became increasingly fragmented.

Spatial theory has also helped to advance research on "heterogeneity" for example. Pickett et al. (2016) see urban systems as mosaics of patch type, which means the types, numbers, boundaries and changes in patches are drivers of system structure and function. This requires mapping green, blue or gray structures in order to better conceptualize the urban system. Here finer spatial resolution data derived from remote sensing platforms is critical (Zhou et al., 2014).

As ecology evolves to integrate social-ecological systems with governance and decision making, geographers will be instrumental in developing spatially explicit tools that for decision makers that can help illuminate overlooked places in the city where policy, planning or community development could enhance biodiversity, ecosystems provisioning and social justice goals (McPhearson et al., 2013).

Geography has a long tradition of accounting for the human impacts on the environment. There is now a body of literature that considers the environmental context of urban life, and in the physical sciences, a growing awareness that cities are environments worthy of study (Benton-Short, Short, & John, 2013; Cronon, 1991; Platt, 2014; Platt, Rowntree, & Muick, 1994). Physical geographers are making important advances in understanding the role of green space, particularly around issues such as water retention, storm water runoff, and biodiversity (Coutts, Tapper, Beringer, Loughnan, & Demuzere, 2013; Lundy & Wade, 2011). In addition, the emergence of geospatial techniques allows geographers to examine green infrastructure and green space in new ways. Finally, political ecology, and more recently urban political ecology has positioned human geographers to make important contributions around urban sustainability by looking at the city as a place in which power is revealed, contested and enforced (Buckley, Boone, & Grove, 2016; Cooke & Lewis, 2010; Gandy, 2008; Swyngedouw, 2004). This literature has revealed the interconnections between the physical and the social, and can be a useful lens through which to examine politics behind green infrastructure distribution and benefits.

Part II: geographic skills and knowledge in green space, green infrastructure and sustainability

As part of a larger research project, we analyzed the sustainability plans for twenty U.S. cities. We selected these to represent a range of diverse population sizes and geographic regions around the U.S.[3] Part of our research focused on determining how aspects of green space are defined, used, measured, and conceptualized. Based on this research, and an analysis of the larger scholarly literature on green space and green infrastructure, we identified four critical areas where geographers can and should contribute to advancing both green infrastructure and green space research and practice. These are: 1) scale, 2) mapping distribution, 3) sensitivity to place and locale and 4) equity and access.

Scale

We first return to the issue of scale relevant to green infrastructure as attention to scale is both central to geography and particularly timely in regards to planning for and research on green amenities. As discussed earlier, definitions of green infrastructure are complicated by scale. Some – including the US EPA – have use this definition to only refer to small-scale stormwater management best practices integrated with the urban fabric while others define it solely in terms of large-scale features like regional park systems (Keeley et al., 2013; EPA, 2014; Weber et al., 2006). Indeed disconnects resulting from stakeholders understanding the term green infrastructure differently has been identified as a barrier to utilization of vacant land and green infrastructure for stormwater management in some Great Lakes cities (Keeley et al., 2013). Our examination of urban sustainability plans similarly reveled that terms like green infrastructure were frequently used without definition yet often assumed some scale within their discussion.

This disconnect can be seen in other ways as well. For instance, landscape ecology concepts such as size, contiguity, proximity, and connectivity (which will be discussed in the next section) have been especially prominent in enhancing the quality of larger-scale green space outside of cities (Baycan-Levent & Nijkamp, 2009). It is only very recently that they have been applied to smaller-scale green space within cities. However, urban ecosystem fragments have been demonstrated to contribute to species preservation more "than their limited size and disturbed state might suggest", making their planning and connectivity of vital importance (Jim, 2004; Rudd, Valal, & Schaefer, 2002).

Concretely, sensitivities to scale dovetail with current research directions in urban ecology. Traditionally, urban ecological research was of two distinct types: one which worked at a relatively small-scale, examining ecological functions of remnant "natural" systems within the urban fabric (Cadenasso & Pickett, 2013), and the second a macro-level analysis of energetics, material budgets, and the metabolism of the city as a whole (Coleman, 2010; Collins et al., 2000; Pataki et al., 2011; Pickett & Cadenasso, 2002). These two schools have been called ecology *in* the city and ecology *of* the city, respectively (Grimm et al., 2008). Contemporary ecologists, often using the tools of geography – such as GIS and remote sensing – have begun to merge these lines of research through efforts to examine cities in their entirety – rather than just vegetated areas – and participate in multidisciplinary efforts to understand cities as integrated social-ecological systems. Significantly, these efforts now attempt to incorporate small-scale physical, biological, and social heterogeneities into their understandings of city-wide processes and changes (Cadenasso, Pickett, & Schwarz, 2007).

It is time for a new green infrastructure paradigm, one that sees green infrastructure elements as unified along a continuum of scale. This conceptual unification could be led by geographers, who are particularly adept at moving up and down scales of analysis. Unification opens new opportunities for exchange and research. This in both illuminating the many similarities that exist between green infrastructure types at all scales and simultaneously highlighting the fact that green amenities offer differentiated functionality dependent on location (Gill et al., 2007, p. 116). It also aligns with trends seen in urban ecology that link studies of small-scale systems and understandings of city-wide processes and changes (Cadenasso et al., 2007). Finally, scale connects with new understandings related to equity as well, with efforts to plan for and understand desired green infrastructure-related benefits not just at the city or neighborhood level, but also for different stakeholders and individuals (Keeley et al., 2013). In each of these realms, geographers are well positioned to help broaden discussions of green infrastructure at across scales and along urban to rural gradients.

Mapping distribution

Many municipal sustainability plans in the U.S. have goals related to the creation or updating of green space or open space master plans and generally seek to increase the area devoted to park and open space. Though simply increasing the quantity of green space is an obvious first step, other green space objectives are also crucial for sustainability planning. For instance, without appropriate attention to the distribution and type of new green space outcomes may not match expectations.

Green space connectivity within cities is emerging as an important goal, and one that contributes not only to habitat value of these systems and utility as transit and recreation corridors for humans but that contributes to system stability and benefits in other ways as well. We see examples of these aspirations in the city of Madison's goal to connect parks, bike trails, and stormwater management systems. Specifically, their aim is to "link all parks and open spaces to the maximum extent possible" (City of Madison, 2011, p. 16).

However, there remain significant barriers to increasing greenspace connectivity. The complexity of the mapping, inventorying, and analysis of existing and potential greenspaces to increase connectivity should not be underestimated. Fortunately, this type of analysis fits with the expertise of geographers who can bring sensitivity to scale and assessment using tools such as GIS and remote sensing. These tools can also assist in connecting green spaces and open spaces within wider planning efforts (Erickson, 2006). For example, in some municipalities, private and unprotected areas can or could constitute an important component of network planning. From large parcels to run-of-the-mill backyards, the extent of private green space viewed cumulatively can be larger than that of public green space (Pauleit, 2003).

Significantly however, many cities have little or no information on private green space, despite its potential importance (Huang et al., 2013). Overarching assessments of green spaces, including privately owned ones are essential for a full understanding of a municipality's green space. The increased availability of parcel-scale imagery and associated geographic techniques will help municipalities address the current information gap and allow better analysis of access, quality and the level of connectivity of natural areas within the metropolitan region.

Mapping the urban tree canopy cover is another challenge faced by US cities. Many factors – not to least their ubiquitousness and decentralized nature make assessments of the distribution of trees a particular challenge. This warrants further discussion here because of the important role that geographic tools such as GIS and remote sensing bring to urban tree canopy planning, distribution and measurement.

Many municipal sustainability plans have green space goals related to tree canopy. New York City's 2007 One Million Trees program to plant 1 million new trees by 2020 is perhaps the most well-known of these goals (City of New York, 2011), although Washington D.C. has a similar goal to "plant 8,600 new trees citywide per year until 2032" (City of Washington DC, 2012, p. 76). Many cities, however, set general goals to increase tree canopy coverage (sometimes by a certain percentage) without including benchmarking or monitoring plans.

Measuring the urban tree canopy – for the purposes of establishing a baseline number of trees or tracking progress towards a set goal – poses particular problems. For example, while municipal departments often try to create inventories of parks and open spaces through field observation, the decentralized nature of street trees means that other methods, including remote sensing and geographic information systems (GIS) are typically more effective (Huang et al., 2013).

Thus, tree canopy is measured quite differently than other forms of green space. Currently, there are two popular and well established technical ways to measure urban tree canopy: 1) using measures of field-based visual estimations with the USDA Forest Services' i-Tree program and 2) the use of GIS and high-resolution land cover data (King & Locke, 2013). Geographers have a particular role to play in refining urban tree measurement techniques based upon remote sensing imagery (including aerial photos and satellite images), which currently have a variety of limitations, briefly outlined here. In a first example, the widely used V-I-S model defines the urban landscape as the sum of Vegetation, Impervious surfaces and Soil, yet does not discriminate well between trees and other forms of vegetation (Ridd, 1995). Indeed, many current methods of estimation are very time consuming and become less accurate with the more types of green space that one attempts to classify (Shin & Lee, 2005). Another problem has been that some methods assume tree cover as a static component of urban areas and insufficiently account for interannual tree cover variability at the city or neighborhood scale (Johnston, 2013; Nowak and Greenfield, 2012).

While a variety of green space and street tree classification systems already exist, these are further limited by differences in green attribute and remote sensing data available in each city and by the expense and time-intensiveness of necessary field work and ground-truthing. As geographers advance and refine the techniques and models used to evaluate green space and tree canopy, their work will assist cities in measuring their existing green spaces and tree canopy, prioritizing and setting more accurate targets and priority areas for increasing coverage. Such developments will have very real and immediate implications. Maps and models of this kind are necessary (though currently lacking) for cities to utilize decentralized small-scale green infrastructure as a form of infrastructure, such as to rely upon this for stormwater management services (Keeley et al., 2013).

Sensitivity to place and locale

Geographers will not be surprised that success can sometimes be determined and even defined by local conditions. Here, we discuss they ways that local environmental conditions can (or should) impact green space planning, the types of vegetation that it contains, and the sustainability of needed maintenance. Not all cities can plan for the same types of parks and open spaces because different geographies and environments call for localization.

For instance, many cities promote the use of native plant species, a measure that can, in some instances, improve quality and make green spaces *more* sustainable. Native species are promoted for many reasons, including habitat value for wildlife and tolerance to the climatic variability of specific regions. For instance, drought tolerance of native species is of particular importance in many regions in the US. While many sustainability plans only vaguely discus of the incorporation of native species, Grand Rapids's plan stands out for its promotion of a specific, yet rational native plants goal: to "increase the percent of low-maintenance and native plants used in landscaping throughout the city by at least 25% by 30 June 2015" (City of Grand Rapids, 2013). In this context however, we should clearly note that in some cases, native plants would not be sustainable options in urban areas due to their intolerance for the modifications that urban areas place upon the environment. For this reason, the specification of low-maintenance and non-invasive species might be more appropriate than a general call for native species.

In fact, in the sustainability plans we examined, we found that other terms were substitutes for "green space." Nearly all of the cities we examined in the Southwest and in California did not use the term "green space". Of course, many of these cities are located in desert environments where the natural vegetation is not necessarily as "green." In fact, the avoidance of "green" in these cities can be viewed as a positive trend because traditional "green" space would likely be energy and water intensive and therefore have a negative and unsustainable connotation within these geographies (Roseland, 2005). Therefore, in these contexts, spaces that are green could actually be unsustainable, such as the lush green lawns of Las Vegas.

A prominent theme in these desert-city plans is a call for more natural and local vegetation. Despite not using the term "green space", cities still reference sub-categories of green space like parks, urban tree canopy, and urban gardens. These cities were more likely to use terms such as "open space" and "natural vegetation". The use of local context (physical, social, economic) is an important factor in sustainability (and especially green space) planning (Baycan-Levent & Nijkamp, 2009; Conroy & Berke, 2004). Research suggests the importance of designing policy instruments for real world conditions rather than trying to make the existing world conform to a particular policy (Tang et al., 2010). This means that local climactic conditions are more important than the popular notion of needing "green" spaces.

Geographers (and some ecologists) contribute to this research by acknowledging the idea of regionally appropriate vegetation, water use, and other maintenance inputs for green space/open space to reach its full sustainable potential. The need for this place sensitivity becomes apparent from our examination of the plans for desert cities where the term "green" can imply unsustainable maintenance of lawns or golf courses. In addition, Berke and Conroy (2000) make the point that sustainability-specific terms like "green space" may not be politically acceptable in some cities. We realize that there are a number of reasons why regions or cities might need to use alternate terms other than green space or green infrastructure. Thus, we would support wider discussion about the ways in which different cities use the term green space to advance a more complete and geographically sensitive definition.

Distributional and procedural equity

Despite equity's prominence in definitions of sustainability, equity is often the least defined and operationalized of the three "E's" and frequently takes a back seat in both planning and politics (Wheeler, 2013). Indeed, scholars have described the lack of inclusion of equity goals and actions generally among U.S. sustainability efforts (Benton-Short et al., 2016; Portney, 2003; Saha, 2009). While cities are increasingly incorporating equity into sustainability plans and indicator projects, the conceptualization and implementation of the concept remains highly constrained. The lack of specific goals related to equity and the sporadic nature with which equity is addressed makes it clear that this topic is not addressed as holistically as other plan elements. Indeed, the ways equity is raised indicated that it may "be primarily symbolic and indicative of [it] ... being a low-priority concern" (Pearsall & Pierce, 2010, p. 571).

Geographers Hamil Pearsall and Joseph Pierce (2010) argue there are two core elements of environmental justice: distributional and procedural justice. Distributional justice refers to the distribution or maldistribution of environmental consequences on traditionally disadvantaged groups in American society (Pearsall & Pierce, 2010). The procedural element refers to the "meaningful involvement" aspect of the definition and focuses on the right of all people to participate in environmental decision-making (Pearsall & Pierce, 2010). Many studies show that information on the access, provision and quality of green space is very limited (Pauleit 2003; Szulczewska & Kaliszuk, 2003). Accordingly, both distributional and procedural inequity challenge green infrastructure planning in US cities and remain topics for further applied geographic research.

Distributional equity

Green space inequity is a complex issue. Literature on the topic of green space inequity often concludes that areas with lower socioeconomic status and/or a higher minority population contain fewer green space resources than their higher socioeconomic and lower minority population counterparts (Vaughan et al., 2013). These sorts of inequities in the amounts and distribution of green space can lead to "reproduced uneven social space" due to the disproportionate distribution of the benefits associated with green spaces (Benton-Short et al., 2013, p. 242). The critical lens of urban political ecology can be used to better understand the political processes behind the placement and enhancement of green space and green infrastructure. For example, there has been much geographical research on water. Matthew Gandy's book *Concrete and Clay: Reworking Nature in New York* examines the relationship between nature, the city and social power with regards to the construction of the city's water supply (Gandy, 2002). Erik Swyngedouw's research has shown that flows of water are deeply bound up with flows of power and influence; water provision in not simply about connecting supply and demand, but about the interconnections between the physical and the social (Swyngedouw, 2004). This line of inquiry positions geographers to make substantial and important contributions around sustainability planning issues such as equity and access, since these are central to examinations. Geographers bring an awareness of understanding the cultural, demographic and development history of a community, thus advancing a better understanding of the wider urban context. We see potential for this critical lens to examine green infrastructure as well.

Indeed, recent studies by geographers have shown that the assumptions about the inequality paradigm of green space distribution are not always correct (Vaughan et al., 2013). Nik Heynen and colleagues found inequitable distribution of urban trees in the US city of Milwaukee (Heynen, Perkins, & Roy, 2006, p. 512) and Christopher Boone and colleagues investigated the distribution of parks in Baltimore, Maryland (Boone, Buckley, Grove, & Sister, 2009). Their research found that a simple look at the distribution of parks shows that African Americans were located closer to parks. However, more in-depth historical research showed a more complex picture. Whites once had lived in such areas, but after white middle-class flight in the 1960s and 1970s African Americans moved into those areas with greater access to parks. Geographic research that combines amenities maps with more in-depth socio-ecological analysis can reveal that inequity (and equity) has deep historical patterns of unequal treatment (Benton-Short et al., 2013).

Our analysis of green space within urban sustainability plans similarly highlights the need for more nuanced, geographical assessments of green space distribution and equity. Most of the cities we studied – even those which prioritized green space increases or improvements – did not explicitly do so in a way that assures equity.

Some cities, however, do include goals and discussions concerning green space and distributional justice, specifically relating to equal access and distribution of parks and open spaces. For instance, among several relevant goals, New York City states that each neighborhood should provide 1.5 acres of open space per thousand people (City of New York, 2011). Rather than aiming for equity, Dunn (2010) specifies that green infrastructure policies should be concentrated in areas to best provide the urban poor with environmental services. Along these lines, Washington D.C.'s plan calls for a unique solution to address accessibility challenges with plans to create small parks and green spaces in areas with inadequate open space as well as investing in "mobile parklets" (City of Washington, DC,

2012). Green space goals measured by neighborhood is another move toward equitable distribution of these resources. Chicago and New York set goals to make sure that every resident in these cities lives "within a 10 minute walk of a park", recreation area or open space" (City of Chicago, 2012, p. 27; City of New York, 2011). This means of measuring green space accessibility is more closely tied to the lived experience of residents, and thus reflects accessibility differentials that could exist between neighborhoods.

Interestingly, tree boxes, gardens, community gardens and other smaller forms of green space are not as commonly associated with distributional equity. Perhaps this is because they are more difficult to map. And, as discussed earlier, green space elements on private property may pose access problems and are beyond the scope of the city services to maintain. Also of note is that community gardens are often discussed primarily in the context of access to "urban food" rather than as green space.

Indeed, consideration of the distribution of green space is further complicated given the multitude of benefits that they provide. Simply, not all green space types provide equal benefits. Table 1 is a list of green space and green infrastructures benefits compiled from

Table 1. The multiple benefits of green space and green infrastructure.

Benefit category	Benefits	Examples of related actions from Sustainability plans*
Environmental	• Urban Heat Island Regulation • Improve Air Quality • Wildlife Habitat/Biodiversity • Stormwater Management • GHG Sequestration • Walter Filtration • Shade Provision	–Revise parking requirements to provide for additional green space (City of Newport News, 2013). –"Increase percent of low-maintenance and native plants used…by city by at least 25% by 30 June 2015" (City of Grand Rapids, 2013) –"plant 8,600 new trees citywide per year until 2032" (City of Washington, DC, 2012, p. 76).
Economic	• Increase Property Values • Economic Development • Energy Savings • Promote Tourism • Jobs and Employment • Use naturally Unbuildable Property	• "incubate innovative new urban agriculture ventures" to support entrepreneurship (City of St. Louis, 2013, p. 73). • identify and support the "use of naturally unbuildable properties… for agricultural use" (City of Madison, 2011, p. 23).
Social	• Promote and Enhance recreation and Physical Activity • Local Health Food • Community Gatherings and Social Interaction • Education • Interaction with Nature/Time spent Outdoors • Beauty and Aesthetics • Promotes Walkability • Sense of Place • Mental and Spiritual Health	• "link all parks and open spaces to the maximum extent possible" (City of Madison, 2011, p. 16). • "coordinate a campaign to encourage citizens to help plant trees and care for the urban forest" and "continue tree planting programs in partnership with community organizations" (City of Austin, 2008, p. 3). • Update the zoning code to allow urban agriculture and gardening broadly defined (City of St. Louis, 2013)

* Please note that most of the actions listed have multiple goals associated with them. We provide them here as examples with the category of benefit they seemed primarily associated with.
Source: by authors based on analysis of twenty U.S. Sustainability Plans.

discussions of these amenities in urban sustainability plans. Green infastructure encompasses a very diverse range of technology and techniques, with some affecting direct users or those immediately adjacent to the amenity, and other benefits having more passive, general, or disparate locational impacts on individuals in a community. It is also important to note that many of the examples in the table provide multiple benefits across the ecological, social and environmental realm. This is an area of recent interest, particularly in terms of public health, and we see this line of inquiry as a tremendous opportunity for geographers.

This discussion of the benefits received by communities via green infrastructure and green space leads to a deeper discussion of green space quality. The quality of green space is important because it reveals factors such as accessibility, safety, amenity provision and type of green infrastructure. While the distance to and number of parks speaks to the spatial access of green space, features such as safety, traffic, and walkability refer to the "social access". For instance, Grand Rapids sets goals for making their parks American Disability Act compliant as part of an attempt to "ensure access to parks and open spaces for all citizens." Distribution and proximity are not the only factors that facilitate or limit green space use and thus examination of green space size and distribution must be expanded to include aspects of quality in order to address issues of equity.

To meet the social, psychological, and equity needs of citizens, green spaces should be easily accessible and be adequate in both quality and quantity (Haq, 2011; Wen, Zhang, Harris, Holtd, & Croft, 2013). Green space quality, including the types of facilities provided and their maintenance and upkeep is yet another factor that impacts "social access" and perceptions of safety. While these issues arise rarely within municipal sustainability plans, there are some oblique references to these issues, including in DC's plan that acknowledges: "In the past, playgrounds, parks and schools were better maintained in wealthier neighborhoods" (City of Washington, DC, 2012, p. 34). Similarly, St. Louis has prioritized inventories of all parks, recreation facilities and open spaces in the city for both distribution and maintenance (City of St. Louis, 2013). Other characteristics of green spaces, such as underutilization or impeded sight lines may contribute to perceptions of a lack of safety. For example, a green space might be intentionally designed with multi-level vegetation (tree canopy, lower level shrubs) to provide better habitat value, but might then be perceived by some in the community as being a safety hazard. Trade-offs clearly exist between some green infrastructure benefits.

Indeed, as Dunn (2010) highlights, while green infrastructure can provide amenities, it also provides dissamenities. In urban environments, these move beyond questions of aesthetic preference and are also safety issues. For instance, green infrastructure can provide habitat for disease vectors, or other undesired species (Dunn, 2010). Indeed, disease prevalence tends to increase as habitats are more managed or disturbed by humans, so the discussion is critical for urban greenspace (Dunn, 2010). Frank conversations about not only green infrastructure amenities but also dissamenities are crucial to provide beneficial and equitable

One of the best examples of the use of geographic analysis we have seen applied to green infrastructure is in the plan for the city of Philadelphia. This plan devoted twenty pages to equity and articulated the goal of achieving more equitable access to healthy neighborhoods (City of Philadelphia, 2009). Philadelphia makes a strong connection between green spaces and the health, specifically the health benefits of reduced asthma and obesity rates (City of

Philadelphia, 2009). The plan is also one of few in our study to include maps. One map shows the open space and parks in the city, with their corresponding ten-minute service area in pink while areas in white are considered underserved (see Figure 1). Areas around Fairmont Park and Pennypack Park in Northeast Philly, Manayunk and Germantown are much better served than many areas in Kensington and North and South Philadelphia. Figure 2 shows the presence and absence of tree cover by neighborhood (City of Philadelphia, 2009, p. 57). While such maps convey important information, only those who know the city well would also know which neighborhoods were traditionally underserved and poor. Thus, in this case, both maps would have benefitted from including data

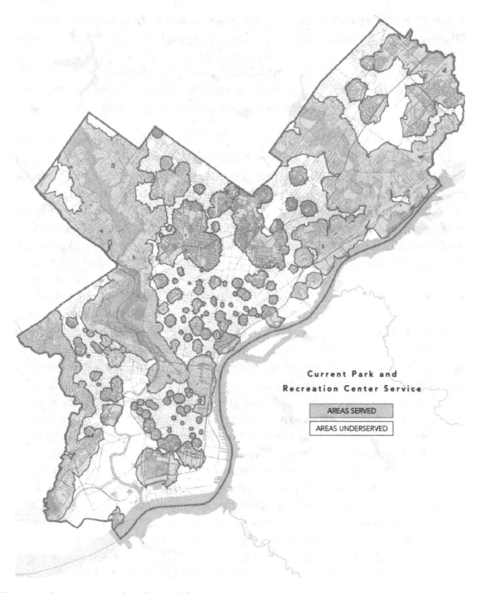

Figure 1. Open space and parks in Philadelphia. Ten-minute walk radius service areas appear light grey; areas in white are considered underserved (City of Philadelphia and WRT Design, 2009, p. 49).

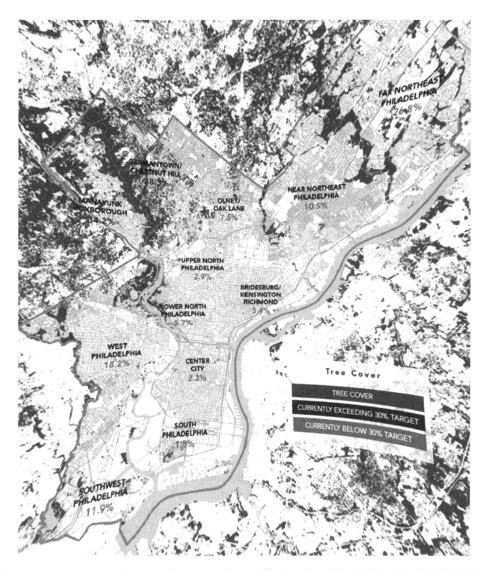

Figure 2. Presence and absence of tree cover by neighborhood in Philadelphia (City of Philadelphia and WRT Design, 2009, p. 57).

such as household income, race and ethnicity, something easily done with GIS. We also note that Philadelphia will also soon develop an online Equity Index, which will identify geographic areas of inequity of sustainability initiatives in the city. This would be an opportunity to highlight such linkages and systematic shortcomings in service provision – something that individual maps (highlighting progress in one specific area) fails to accomplish.

Equity in process
Geographers used to working in urban development at the neighborhood or local scale are attuned to the importance of process. Green space improvement or creation and should include deliberative and inclusionary processes and procedures to actively

engage the public in decision making related to green space location, desired amenities or safety (Seymour, 2012). Such community involvement in revitalization projects serve an important role in ensuring community needs are met and that new or improved "spaces are not perceived as inferior", unwanted or unsafe (Wendel, Heather, Downs, & Mihelcic, 2011, p. 6731). For instance, communities with different age, gender or ethnic compositions may wish to tailor their green space plans to local needs and preferences (Payne, 2002). However, many cities sustainability plans do not have explicit goals to include local residents in decision-making about desired amenities or safety related to green space and green infrastructure.

Lessons could perhaps be drawn from efforts to engage residents in care for street trees, which was one community-engagement effort featured in a number of sustainability plans. For example, Austin has two related goals: first, to "coordinate a campaign to encourage citizens to help plant trees and care for the urban forest" and second to "continue tree planting programs in partnership with community organizations" (City of Austin, 2008, p. 3). Such efforts could help engage residents with horticulture and sustainability, foster a connection with nature and the local environment, and provide an avenue for multilateral communication about local needs, priorities and values.

The cultivation of partnerships is also important, because in many cases, non-profit organizations and community groups are in charge of local programs. In Washington, D.C., for example, the non-profit Casey Trees is primarily responsible for all public space tree plantings and work directly with the community to organize tree planting events. They also coordinate a planting assistance program for residents planting trees on private property. In addition, this non-profit has as a GIS expert on staff and tree canopy research conducted by this organization was used by the city to develop municipal goals to increase the tree canopy. In this way, community partners can provide information about safety, traffic, walkability at a local scale.

There remains much room for work on participatory equity as there is still a gulf between the frequency with which equity is "mentioned and the depth with which it is explored" (Steele, Diana., Byrne, & Houston, 2012, p. 77). Cities, in practice, and researchers as well need to move beyond city-wide goals and analysis, and to think about access and amenity quality and accessibility at a smaller scales such as neighborhoods and wards, or even to individuals. Geographers may, for instance, be able to provide historic and contemporary perspectives on local context and community needs (Pauleit 2003). This is an area where geographers can clearly contribute to discussions and develop best practices moving forward. Such critical analysis can also help assess the level of community participation in sustainability efforts.

Another important contribution from Geography with regards to equity and access is the tool of GIS to assist with geographic analysis in both distributional and procedural equity. Consider the example of "food deserts" a term that refers to areas of the city with limited access to fresh and affordable food. Geographer Neil Wrigley's work on food deserts in British cities shows that these deserts reflect social exclusions and contribute to health inequalities (Wrigley, 2002). Work on food deserts has continued to advance; the tools and the approaches used can be employed to examine a range of equity issues. Geographers who employ such approaches also allow us to re-envision the city as more equitable (Haughton, 1999; Stefanovic & Scharper, 2012).

GIS has tremendous potential to inform analysis around green space and green infrastructure distribution and equity. It also allows planners the ability to prioritize interventions and expand on goals based on the power of visualizing disparity (Bell, Montarzino, & Travlou, 2007). A combination of geographic techniques and critical analysis allows us to investigate historical land use patterns, examine policies around parks and recreation, and the politics of funding of public amenities. These are important roles in the larger picture of sustainability planning

Conclusions

Geographers are well skilled in examining complex urban natural and social issues, and can contribute to the larger discussion of urban sustainability by applying the tools of geography to green infrastructure research. We are not the first to make this observation, but here we draw attention to the important research collaborations needed on topics of green infrastructure and green space. This is because it provides so many potential benefits and requires the work of multiple disciplines to be fully valued. Common areas of interest, particularly around socio-ecological systems, make interdisciplinary work on green infrastructure an important part of sustainable urban planning. Within this broad review of green infrastructure theory and practice, we identified four areas in which geography can offer unique contributions.

First, we identified the need for a unified conceptualization of green infrastructure elements along a continuum of scale. The current disconnect has resulted in the development of entire research methodologies and theoretical constructs applied to green infrastructure of only large *or* small scales. This disconnect has – in part – been necessary because the complexity of urban systems was too great to incorporate detailed information at neighborhood or parcel-levels. However, geographic sensitivity to scalar work generally, and spatial analysis and modeling skills particularly are greatly needed as the field moves forward in this regard.

Second, we draw attention to both the theoretical and practical challenges associated with mapping the distribution of green spaces in urban areas. We see particular opportunities for geographers in using mapping techniques to understand the benefits of different types of green infrastructure and those on public as well as private properties. Specific and applied challenges exist in planning for increased green space connectivity and in understating the social, economic and environmental benefits of diffuse, small-scale green infrastructure.

Third, we see a key role of geographers in a place-based understanding of green infrastructure and its role in sustainability. A lush, vegetated landscape might contribute positively to sustainability efforts in one city and negatively in another. We found that even the term "green space" is, rightly, not unified across ecological regions. As new concepts and understandings about green space emerge, these efforts must be tempered with a sensitivity to local environmental conditions and resulting regional differences in vegetation types that make green spaces contributors to urban sustainability.

Finally, we highlight the roles that geospatial techniques and critical geography can play in assessments of green space equity. Attention to procedural equity will assure effective inclusionary decision making processes such that all segments of the population have input concerning green space location, desired amenities or safety. Roles for geographers in considering distributional equity ranges include technical applications of geospatial technique and nuanced understandings of the differing benefits associated with myriad types of green infrastructure (trees, wetlands, green roofs) and how these are distributed across the landscape. In addition, a geographer's skills in both critical analysis and geospatial technology are important tools in uncovering the types of multiple co-benefits that green infrastructure can provide.

Research in urban sustainability requires a synergy among interdisciplinary scholars. The study of green infrastructure is at a particularly exciting moment. With its "rebranding" as green infrastructure, vegetation has newly found a new and central role in sustainability efforts, given its potential to address all three pillars of sustainability: environment, economy, and equity. Geographers possess a sensitivity to scale, a critical perspective on power and place, and technical skills that that is crucial to furthering green infrastructure research at the urban environmental nexus.

Notes

1. It should be noted that there is also no global standard either, and as a result sustainability plans are varied and un-standardized documents (ICELI, 2008).
2. Ecologists have tended not to focus on cities, and this disconnect is reflected in the fact that today the LTER program includes only 2 cities out of its 26 research locations.
3. We selected these twenty cities to research based on three criteria. First, to be included in the study, a city must have a holistic sustainability plan that was formally approved by city council, included a publication date, and existed in a downloadable format as of 2013. Second, plans also had to have a primary focus on sustainability. In this regard, preference was given to plans that used the term sustainability in their title or description; however, three climate action plans were included because they have a sufficiently comprehensive approach to sustainability. Third, we selected a range of cities based on population size and geography.

Disclosure statement

No potential conflict of interest was reported by the authors.

References

Baycan-Levent, Tü.zin, & Nijkamp, Peter. (2009). Planning and management of urban green spaces in Europe: Comparative analysis. *Journal of Urban Planning and Development, 135*(1), 1–12.

Bell, Simon, Montarzino, Alicia, & Travlou, Penny. (2007). Mapping research priorities for green and public urban space in the UK. *Journal of Urban Forestry and Greening, 6,* 103–115.

Benton-Short, Lisa, Keeley, Melissa, & Rowland, Jennifer (2016). "Green space in urban sustainability plans: trends and best practices" K. Archer & K. Bezdecny (Ed.), *International Handbook of Cities and the Environment.* Edward Elgar. (pp 279–315). http://www.e-elgar.com/shop/handbook-of-cities-and-the-environment

Benton-Short, Lisa, Short, J. R., & John, R. (2013). *Cities and nature* (2nd ed.). New York: Routledge.

Berke, Philip R., & Conroy, Maria. Manta. (2000). Are we planning for sustainable development? An evaluation of 30 comprehensive plans. *Journal of the American Planning Association, 66*(1), 21–33.

Bolund, Per, & Hunhammar, Sven. (1999). Ecosystem services in urban areas. *Ecological Economics. 29*, 293–301. Retrieved from http://www.fao.org/uploads/media/Ecosystem_services_in_urban_areas.pdf

Boone, Christopher G., Buckley, Geoffrey L., Grove, J. Morgan, & Sister, Chona. (2009). Parks and people: An environmental justice inquiry in Baltimore, Maryland. *Annals of the Association of American Geographers, 99*, 767–787.

Boone, Christopher, & Moddares, Ali. (2006). *City and environment.* Philadelphia, PA: Temple University Press.

Buckley, Geoffrey L., Boone, Christopher G., & Grove, J. Morgan. (2016). The greening of Baltimore's Asphalt schoolyards. *Geographical Review*, 1–20.

Cadenasso, Mary L., Pickett, Stewart T. A., & Schwarz, Kristen. (2007). Spatial heterogeneity in urban ecosystems: Reconceptualizing land cover and a framework for classification. *Frontiers of Ecology and the Environment, 5*, 80–88.

Cadenasso, Mary L., & Pickett, Stewart. T. A. (2013). Three tides: The development and state of the art of urban ecological science. In Stewart. T. A. Pickett, Mary L. Cadenasso, Brian McGrath, et al. (Eds.), *Resilience in ecology and urban design: Linking theory and practice for sustainable cities* (pp. 29–46). Dordrecht: Springer Science++Business Media Dordrecht.

City of Austin. (2008). *Rethink Austin.* [Brochure] Retrieved from http://austintexas.gov/sites/default/files/files/Sustainability/Rethink_-_SAA/Rethink_%26_Sustainability_Action_Agenda.pdf

City of Chicago. (2012). 2015 sustainable chicago action Agenda. Retrieved from http://www.cityofchicago.org/content/dam/city/progs/env/SustainableChicago2015.pdf

City of Grand Rapids. (2013). Sustainability plan. (as Amended 2013). Retrieved from http://grcity.us/enterprise-services/officeofenergyandsustainability/Documents/2013%20Amended%20Sustainability%20Plan.pdf

City of Madison. (2011). The Madison sustainability plan: Fostering environmental, economic and social resilience. Retrieved from http://www.cityofmadison.com/sustainability/documents/SustainPlan2011.pdf

City of New York. (2011). PlaNYC: A greener, greater New York. Retrieved from http://www.nyc.gov/html/planyc/downloads/pdf/publications/planyc_2011_planyc_full_report.pdf

City of Newport News (2013), City of newport news roadmap to sustainability, accessed 18 august 2015 at http://www.nngov.com/DocumentCenter/View/1586.

City of Philadelphia. (2009). Greenworks Philadelphia. Retrieved from http://www.phila.gov/green/greenworks/pdf/Greenworks_OnlinePDF_FINAL.pdf

City of St. Louis. (Adopted 2013). City of St. Louis sustainability plan. City of St. Louis Planning Commission. Retrieved from https://www.stlouis-mo.gov/government/departments/mayor/documents/upload/STL-Sustainability-Plan.pdf on 8/18/2015

City of Washington, DC. (Adopted 2012). Sustainable DC plan. Retrieved from http://sustainable.dc.gov/sites/default/files/dc/sites/sustainable/page_content/attachments/DCS-008%20Report%20508.3j.pdf

Coleman, David C. (2010). *Big ecology: The emergence of ecosystem science.* Berkeley CA: University of California Press.

Collins, James P., Kinzig, Ann, Grimm, Nancy B., Fagan, William F., Hope, Diane., Jianguo, Wu, & Borer, Elizabeth T. (2000). A new urban ecology: Modeling human communities as integral parts of ecosystems poses special problems. *American Scientist. 88*, 416–425. Retrieved from www.jstor.org/stable/27858089

Conroy, Maria Manta, & Berke, Philip R. (2004). What makes a good sustainable development plan? An analysis of factors that influence principles of sustainable development. *Environment and Planning A, 36*, 1381–1396.

Cooke, Jason, & Lewis, Robert. (2010). The nature of circulation: The urban political ecology of Chicago's Michigan avenue bridge, 1909-1930. *Urban Geography, 31*, 348–368.

Coutts, Andrew M., Tapper, Nigel J., Beringer, Jason L., Loughnan, Margaret, & Demuzere, Matthias. (2013). Watering our cities: The capacity for water sensitive urban design to support urban cooling and improve human thermal comfort in the Australian context. *Progress in Physical Geography, 37*, 2–28. doi:10.1177/0309133312461032

Cronon, William. (1991). *Nature's metropolis: Chicago and the great west*. New York: Norton.

Dunn, Alexandra Dapolito. (2010). Siting green infrastructure: Legal and policy solutions to alleviate urban poverty and promote healthy communities. *Boston College Environmental Affairs Law Review. 37*, 41–66. Retrieved from http://lawdigitalcommons.bc.edu/ealr/vol37/iss1/3

Erickson, Donna. (2006). *Metrogreen: Connecting open space in North American cities*. Washington, DC: Island Press.

Evenson, Kelly R., Aytur, Semra, Daniel AS., Rodriguez, & Salvesen, David. (2009). Involvement of parks and recreation professionals in pedestrian plans. *Journal of Park and Recreation Administration, 27*(3), 132–142. Retrieved from http://scholars.unh.edu/hmp_facpub/8

Foster, Josh, Lowe, Ashley, & Winkelman, S., (2011). The Value of Green Infrastrcucture for Urban Climate Change Adaptation. Retrieved from Center for Clean Air Policy:. Access September 25, 2016 http://dev.cakex.org/sites/default/files/Green_Infrastructure_FINAL.pdf

Gagné, Sara A., Eigenbrod, Felix, Bert, Daniel G., Cunnington, Glenn M., Olson, Leif T., Smith, Adam C., & Fahrig, Lenore. (2015). A simple landscape design framework for biodiversity conservation. *Landscape and Urban Planning, 136*, 13–27. doi:10.1016/j.landurbplan.2014.11.006

Gandy, Matthew. (2002). *Concrete and clay: Reworking nature in New York city*. Cambridge, MA: MIT Press.

Gandy, Matthew. (2008). Landscapes of disaster: Water, modernity and urban fragmentation in Mumbai. *Environment and Planning A, 40*, 108–130. doi:10.1068/a3994

Gill, Susannah E., Handley, John F., Ennos, A Roland, & Pauleit, Stephan. (2007). Adapting cities for climate change: The role of the green infrastructure. *Built Environment, 33*, 115–133. doi:10.2148/benv.33.1.115

Grimm, Nancy B., Faeth, S. H., Golubiewski, N. E., Redman, C. L., Wu, J., Bai, X., & Briggs, J. M. (2008). Global change and the ecology of cities. *Science, 319*(5864), 756–760. doi:10.1126/science.1150195

Haq, Shah Md. Atiqul. (2011). Urban green spaces and an integrative approach to sustainable environment. *Journal of Environmental Protection, 2*, 601–608. doi:10.1201/b18713-10

Haughton, Graham. (1999). Environmental justice and the sustainable city. *Journal of Planning Education and Research, 18*, 233–243. doi:10.1177/0739456X9901800305

Heynen, Nik, Perkins, Harold. A., & Roy, Parama. (2006). The ppolitical eecology of uuneven uurban ggreen sspace: The impact of political economy on race and ethnicity in producing environmental inequality in milwaukee. *Urban Affairs Review, 42*, 3–25.

Horwood, K. Aren. (2011). Green infrastructure: Reconciling urban green space and regional economic development: Lessons learnt from experience in England's north-west region. *Local Environment, 16*(10), 963–975. doi:10.1080/13549839.2011.607157

Huang, Yan, Yu, Bailang, Zhou, Janhuan, Hu, Chunlin, Tan, Wenqi, Hu, Zhiming, & Wu, Jianping. (2013). Toward automatic estimation of urban green volume using airborne LiDAR data and high resolution Remote Sensing images. *Frontiers of Earth ScienceHigher Education Press and Springer-Verlag, 7*(1), 43–54. doi:10.1007/s11707-012-0339-6

ICLEI. (2008). Local governments for sustainability USA. Retrieved from http://www.icleiusa.org/

Ignatieva, Maria, Stewart, Glenn. H., & Meurk, Colin. (2011). Planning and design of eecological nnetworks in uurban aareas. *Landscape Ecological Engineering, 7*, 17–25. doi:10.1007/s11355-010-0143-y

Jim, Chi Yung. (2004). Green-space preservation and allocation for sustainable greening of compact cities. *Cities, 21*(4), 311–320. doi:10.1016/j.cities.2004.04.004

Johnston, Andrew. (2013). *Tree cover variability in the district of Columbia*(Unpublished Doctoral Dissertation). University of Maryland, College Park, MD

Keeley, M., Koburger, A., Dolowitz, David P. M., D., Nickel, D., William, S... (2013). Perspectives on the use of green infrastructure for stormwater management in Cleveland and Milwaukee. *Environmental Management, 51*(6), 1093–1108. doi:10.1007/s00267-013-0032-x

King, Kristen L., & Locke, Dexter H. (2013). A comparison of three methods for measuring local urban tree canopy cover. *Arboriculture and Urban Forestry*, *39*(2), 62–67. Retrieved from https://www.nrs.fs.fed.us/pubs/jrnl/2013/nrs_2013_king-k_001.pdf

Lundy, Lian, & Wade, Rebecca. (2011). Integrating sciences to sustain urban ecosystem services. *Progress in Physical Geography*, *35*(5), 653–669. doi:10.1177/0309133311422464

Mark, Benedict & Eric. McMahon. (2006). *Green infrastructure: Linking landscapes and communities*. Washington, DC: Island Press.

McDonnell, Mark. (2015). Editorial: Linking and promoting research and practice in the evolving discipline of urban ecology. *Journal of Urban Ecology*, *1*(juv003). doi:10.1093/jue/juv003

McPhearson, Timon, Kremer, Peleg, & Hamstead, Zoe A. (2013). Mapping ecosystem services in New York City: Applying a social-ecological approach in urban vacant land. *Ecosystem Services*, *5*, 11–26. doi:10.1016/j.ecoser.2013.06.005

Mell, Ian C. (2012). Green infrastructure planning: A contemporary approach for innovative interventions in urban landscape management. *Journal of Biourbanism*. *1*, 29–39. Retrieved from http://www.academia.edu/1276378/Green_Infrastructure_planning_A_contemporary_approach_for_innovative_interventions_in_urban_landscape_management

Montalto, Franco, Behr, Christopher, Alfredo, Katherine Ann, Wolf, Max, Arye, Matvey, & Walsh, Mary. (2007). Rapid assessment of the cost-effectiveness of low impact development for CSO control. *Landscape and Urban Planning*, *82*(3), 117–131. doi:10.1016/j.landurbplan.2007.02.004

Netusil, Noelwah R., Levin, Zachary, Shandas, Vivek, & Hart, Ted. (2014). Valuing green infrastructure in Portland, Oregon. *Landscape and Urban Planning*, *124*, 14–21. doi:10.1016/j.landurbplan.2014.01.002

Nowak David, J., & Greenfield, Eric J. (2012). Tree and impervious cover change in U.S. cities. *Urban Forestry and Urban Greening*, *11*, 21–30. doi:10.1016/j.ufug.2011.11.005

Parker, Paul, & Rowlands, Ian H. (2007). City partners maintain climate change action despite national cuts: Residential energy efficiency programme valued at local level. *Local Environment*, *12*(5), 505–517. doi:10.1080/13549830701656853

Pataki, Diane E., Carreiro, Margaret. M., Cherrier, Jennifer., Grulke, Nancy. E., Jennings, Viviece., Pincetl, Stephanie., ... Zipperer, Wayne. C. (2011). Coupling biogeochemical cycles in urban environments: Ecosystem services, green solutions, and misconceptions. *Frontiers in Ecology and the Environment*, *9*, 27–36. doi:10.1890/090220

Pauleit, Stephan. (2003). Perspectives on urban greenspace in Europe. *Built Environment*, *29*(2), :89-93. doi:10.2148/benv.29.2.89.54470

Payne, Karen. (2002). Graph theory and open-space network design. *Landscape Research*, *27*(2), 167–179. doi:10.1080/01426390220128640

Pearsall, Hamil, & Pierce, Joseph. (2010). Urban sustainability and environmental justice: Evaluating the linkages in public planning/policy discourse. *Local Environment*, *15*(6), 569–580. doi:10.1080/13549839.2010.487528

Pickett, Stewart T., Cadenasso, A., Mary, L., Childers, Daniel L., McDonnell, M., & Zhou, W. (2016). Evolution and future of urban ecological science: Ecology *in, of,* and *for* the city. *Ecosystem Health and Sustainability*, *2*(7), e01229. doi:10.1002/ehs2.1229

Platt, Rutherford H. (2014). *Reclaiming American cities: Tthe struggle for people, place and nature since 1900*. Amherst: University of Massachusetts Press.

Platt, Rutherford H., Rowntree, Rowan A., & Muick, Pamela C. (Eds). (1994). *The ecological city: Preserving and restoring urban biodiversity*. Amherst: University of Massachusetts Press.

Portney, Kent H. (2003). *Taking sustainable cities seriously; economic development, the environment and quality of life in American cities*. Cambridge, MA: The MIT Press Cambridge, MA.

Ridd, Merrill K. (1995). Exploring a V-I-S (vegetation-impervious surface-soil) model for urban ecosystem analysis through remote sensing: Comparative anatomy for cities. *International Journal of Remote Sensing*, *16*(12), 2165–2185. doi:10.1080/01431169508954549

Robbins, Paul. (2012). *Political ecology: A critical introduction* (2nd ed., pp. 2edn). Chichester, West Sussex: Wiley-Blackwell.

Roseland, Mark. (2005). *Toward sustainable communities: Resources for citizens and their governments* (Revised Edition). Gabriola Island, BC: New Societies.

Rudd, Hillary, Valal, Jamie, & Schaefer, Valentin. (2002). Importance of backyard habitat in a comprehensive biodiversity conservation strategy: A connectivity analysis of urban green spaces. *Restoration Ecology, 10*(2), 368–375. doi:10.1046/j.1526-100X.2002.02041.x

Saha, Davashree. (2009). Empirical research on local government sustainability efforts in the USA: Gaps in the current literature. *Local Environment, 14*(1), 17–30. doi:10.1080/13549830802522418

Schäffler, Alexis, & Swilling, M. (2013). Valuing green infrastrcucture in an urban environment under pressure-The Johannesburg case. *Ecological Economics, 86*, 246–257. doi:10.1016/j.ecolecon.2012.05.008

Schilling, Joseph, & Logan, Jonathan. (2008). Greening the rust belt: A green infrastructure model for right sizing America's shrinking cities. *JapaAPA, 74*(4), 451–466. doi:10.1080/0194436080235495610.1080/01944360802354956

Seto, Karen C., Güneralp, Burak, & Hutrya, Lucy R. (2012). Global forecasts of urban expansion to 2030 and direct impacts on biodiversity and carbon pools. *Proceedings of the National Academy of Sciences of the United States of America, 109*(40), 16083–16088. doi:10.1073/pnas.1211658109

Seymour, Mona. (2012). Just sustainability in Urban Parks. *Local Environment, 17*((2),), 167–185. doi:10.1080/13549839.2011.646968

Shin, Dong-hoon, & Lee, Kyoo-seock. (2005). Use of remote sensing and geographical information systems to estimate green space surface-temperature change as a result of urban expansion. *Landscape Ecology Engineering, 1*, 169–176. doi:10.1007/s11355-005-0021-1

Short, John R. (1989). Cities as if only some people matter. In *The humane city*. Oxford: Blackwell.

Steele, Wendy, Diana., Maccallum, Byrne, Jason., & Houston, D. (2012). Planning the Climate-just Cityty. *International Planning Studies, 17*(1), 67–83. doi:10.1080/13563475.2011.638188

Stefanovic, Ingrid LemanG., & Scharper, Stephen. Bede. (2012). *The natural city: Re-envisioning the built environment*. Toronto: University of Toronto Press.

Swyngedouw, Erik. (2004). *Social power and the urbanization of Water*. Oxford: Oxford University Press.

Szulczewska, Barbara, & Kaliszuk, Ewa. (2003). Challenges in the planning and management of 'Greenstructure' in Warsaw, Poland. *Built Environment, 29*(2), 144–156. Retrieved from http://www.jstor.org/stable/23288813

Tang, Zhenghong, Brody, S., Samuel, D., Quinn, C., Courtney, L., Chang, Liang, & Wei, T. (2010). Moving from agenda to action: Evaluating local climate change action plans. *Journal of Environmental Planning and Management, 53*(1), 41–62. doi: 10.1080/09640560903399772

Tzoulas, Konstantinos, Korpela, Kalevi, Venn, Stephen, Yli-Pelkonen, Vesa, Kazmierczak, Aleksandra, Niemela, Jari, & James, Philip. (2007). Promoting ecosystem and human health in urban areas using green infrastructure: A literature review. *Landscape and Urban Planning, 81*, 167–178. doi:10.1016/j.landurbplan.2007.02.001

United States Environmental Protection Agency (US EPA). (2014). What is green infrastructure? Retrieved from http://water.epa.gov/infrastructure/greeninfrastructure/gi_what.cfm on 8/16/2015

Vandermueluen, Valerie, Verspecht, Ann., Vermeire, Bert., Van Huylenbroeck, Guido., & Gellynck, X. (2011). The use of economic valuation to create public support for green infrastructure investments in urban areas. *Landscape and Urban Planning, 103*(2), 198–206. Retrieved from https://www.sciencedirect.com/science/article/pii/S0169204611002428

Vaughan, Katherine B., Kaczynski, Andrew T., Stanis, Wilehlm, Sonja, A., Besenyi, Gina M., Bergstron, Ryan, & Heinrich, Katie M. (2013). Exploring the distribution of park availability, features and quality across Kansas City, Missouri by Income and Race/Ethnicity: An environmental justice investigation. *Annals of Behavioral Medicine, 45*, S28– 38. doi:10.1007/s12160-012-9425-y

Weber, Ted, Sloan, Anne, & Wolf, John. (2006). Maryland's green infrastructure assessment: Development of a comprehensive approach to land conservation. *Landscape and Urban Planning, 77,* 94–110. doi:10.1016/j.landurbplan.2005.02.002

Wen, Ming., Zhang, X.ingyou, Harris, Carmen D., Holtd, James B., & Croft, Janet B. (2013). Spatial disparities in the distribution of parks and green spaces in the USA. *Annals of Behavioral Medicine, 45*(Suppl 1), S18–S27. doi:10.1007/s12160-012-9426-x

Wendel, Wright, Heather, E., Downs, Joni A., & Mihelcic, James R. (2011). Assessing equitable access to urban green space: The role of engineered water infrastructure. *Environmental Science and Technology, 45,* 6728–6734. doi:10.1021/es103949f

Wheeler, Stephen M. (2013). *Planning for sustainability: Creating livable, equitable, and ecological communities* (2nd ed.). New York, NY: Routledge Press.

Wrigley, Neil. (2002). 'Food Ddeserts' in British cities: Policy context and research practice. *Urban Studies, 39,* 2029–2040. doi:10.1080/0042098022000011344

Young, Robert, Zanders, Julie, Lieberknecht, Katherine, & Fassman-Beck, Elizabeth. (2014). A comprehensive typology for mainstreaming urban green infrastructure. *Journal of Hydrology, 519*(November), 2571–2583. doi:10.1016/j.jhydrol.2014.05.048

Zhou, Weiqi, Cadenasso, Mary L., Schwarz, Kirsten, & Pickett, Stephen T. A. (2014). Quantifying spatial heterogeneity in urban landscapes: Integrating visual interpretation and object-based classification. *Remote Sensing, 6,* 3369–3386. doi:10.3390/rs6043369

Zhou, Weiqi, Huang, Ganlin, Pickett, Stephen T., & Cadenasso, Mary. (2011). 90 years of forest cover change in the urbanizing Gwynns Falls watershed, Baltimore, Maryland: Spatial and temporal dynamics. *Landscape Ecology, 26,* 645–659. doi:10.1007/s10980-011-9589-z

Uneven urban metabolisms: toward an integrative (ex)urban political ecology of sustainability in and around the city

Innisfree McKinnon, Patrick T Hurley, Colleen C Myles, Megan Maccaroni and Trina Filan

ABSTRACT
Expanding cities present a sustainability challenge, as the uneven proliferation of hybrid landscape types becomes a major feature of 21st century urbanization. To fully address this challenge, scholars must consider the broad range of land uses that being produced beyond the urban core and how land use patterns in one location may be tied to patterns in other locations. Diverse threads within political ecology provide useful insights into the dynamics that produce uneven urbanization. Specifically, urban political ecology (UPE) details how economic power influences the development decision-making that proliferate urban forms, patterns of uneven access, and modes of decision-making, frequently viewing resource extraction and development through the urban metabolism lens. The political ecology of exurbia, or, perhaps, an *exurban* political ecology (ExPE), examines the symbolic role nature and the rural have played in conservation and development efforts that produce social, economic, and environmental conflicts. While UPE approaches tend to privilege macroscale dynamics, ExPE emphasizes the role of landowners, managers, and other actors in struggles over the production of exurban space, including through decision-making institutions and within the context of broader political economic forces. Three case studies illustrate the strengths and weaknesses of these approaches, demonstrating the benefits for and giving suggestions on how to integrate their insights into urban sustainability research. Integrated political ecology approaches demonstrate how political-economic processes at a variety of scales produce diverse local sustainability responses.

Introduction

We live in an urban age. Yet despite the variety of social and environmental impacts of rapid urbanization around the globe (e.g., loss of agricultural lands, deforestation, increased threats to biodiversity), exactly what it means to live in an "urban age" remains unclear. Moreover, understanding which spaces are urban and which "remain" rural is also unclear. Indeed, large percentages of the "urban" population live in spaces still often overlooked by urban geography: in sprawling suburbs, edge cities, exurbs, informal settlements, and small cities and towns. Individuals in these extra-urban places inhabit, are integrated into, and interact with urban places and networks. Yet often discussions of urban sustainability in both academic and popular literature focus on sustainability in the urban core and problematically ignore these spaces of extended urbanization (Brenner & Schmid, 2014).

A focus on the city proper or urban core within studies of urban sustainability is problematic for both theoretical and empirical reasons. Lefebvre's theory of planetary urbanization predicts that over time the separation between urban and rural spaces will become less and less distinct. And indeed, this mixing of urban and rural has been noted by numerous scholars (Dirksmeier, 2012; Furuseth & Lapping, 1999; Hiner, 2014; Jansson, 2013; Lacour & Puissant, 2007; Olson & Munroe, 2012; Sandoval & Maldonado, 2012). On an empirical level it is impractical to create an artificial divide between the circulation of energy and materials within urban centers and their arrival and eventual disposal outside the city proper. Despite recent attention to urban agriculture, cities will never be materially self-sufficient, but rather will continue to rely on rural spaces at the source of basic necessities, including food, water, energy, and building materials. As such, research in urban sustainability which lacks attention to the exchange of material and energy between urban and rural spaces is sadly incomplete. In this article, we propose that work by political ecologists can contribute to an expanded focus for urban sustainability, enriching the field through a focus on sustainability politics broadly imagined, both formal and informal, local and regional.

Political ecology approaches offer several key insights to research on urban sustainability. First, political ecologists ask "sustainability for whom"? Whose visions and idea (l)s of sustainability are being enacted? As political ecologists, we are reticent to attempt to define sustainability, preferring to focus our research on issues related to access to land and other natural resources and the economic, political, and social power dynamics invoked by stakeholders as they enact their visions on the landscape. Second, as recent work in urban theory has pointed out, urbanization processes produce uneven results in particular places including rapid gentrification, deindustrialization, inadequate informal housing, suburbanization, exurbanization, and the restructuring of rural places and economies. Political ecology, with its careful attention to particular places and processes of power, combined with awareness of the broad, multiscalar processes at work, provides insights into how political and economic power enables or constrains a range of actors in enacting their visions of sustainability.

In order to examine how political ecology contributes to work on urban sustainability, we must acknowledge that the work on urbanization by political ecologists has tended to be divided into two somewhat separate literatures: urban political ecology and a political ecology of exurbia. Political ecology approaches emerged from international research on land tenure, rights, and management in the developing world. Because of that history, it took particular scholars turning our attention to similar issues and dilemmas "at home" for political ecologists to begin working in domestic contexts (Fortmann, 1996). What developed after that, after a brief discussion of "first" versus "third" world political ecology, was a literature largely focused on how traditionally rural places were being impacted by the uneven processes of urbanization (P. Hurley & Taylor, 2016; Reed, 2007; Walker & Fortmann, 2003). This "political ecology of exurbia" (hereafter exurban political ecology) is not an attempt to coin a new subfield, but rather a recognition of the need within political ecology more widely to acknowledge work already being done across various literatures. Urban political ecology emerges later with a call for urban studies to re-nature urban spaces, bringing the insights of political ecology to yet another space, the urban core (Heynen, Perkins, & Roy, 2006) – and only more recently turning its focus to spaces of what has been called "extended urbanization".

We argue in this paper that these two political ecology approaches to urbanization provide valuable insights into the social, economic, and political processes at work. Moreover, we suggest that combining them strengthens our perspectives and analytical purchase on uneven spaces of urbanization. We note that while Brenner and Schmid 2013 have characterized spaces of "extended urbanization" as low density, sprawling settlement, others have argued that this growing phenomenon could also be characterized as extended ruralization (Krause, 2013; Mercer, 2016). Exurbia, or periurban spaces more broadly, in many ways represent a meeting or overlapping of dynamics associated with the urban and the rural, a distinct and emergent landscape in-between (Taylor & Hurley, 2016). We use the comparative approach suggested by Taylor and Hurley (2016) to briefly examine three case studies that endeavor to unravel how exurban and urban political ecology approaches might be productively integrated to produce a more complete picture of the socioecological changes taking place in extended megapolitian regions (Gottmann, 1957; Gustafson, Nik Heynen, Rice, Marshall Shepherd, & Strother, 2014). The paper begins by reviewing research in urban and exurban political ecology, outlining how these two literatures have tended engage with (ex)urbanization. Then we examine each case to illustrate how these two approaches might productively be combined. We conclude with a discussion of what political ecology approaches can offer discussions of urban sustainability.

Extended Urbanization and Integrated (Ex)Urban Political Ecology

As has been noted elsewhere, political ecology has become a sprawling interdisciplinary literature encompassing a multitude of different approaches (see e.g., Robbins, 2012; Watts & Peet, 2004). Our intention here is not to give a comprehensive overview, but rather to focus on how political ecologists have approached the study of urbanization in the Global North. Political ecology in North America and the Global North more broadly has emerged from two separate moments of engagement with other literatures. This has led to the development of a split literature largely separated by geography but also tending to approach land use change with somewhat different foci. As Blaikie (2008) pointed out, sometimes disjunctures come about not so much because of unresolved debates, but rather because of non-engagement (see also McKinnon & Hiner, 2016). In this way, we find two threads of (ex)urban political ecology: urban political ecology and the political ecology of exurbia.

Urban political ecology has focused largely on socio-environmental issues *within cities proper*, framing the ecological impacts and power differentials driving them using the concept of the metabolism of nature (Keil, 2005). Exurban political ecology has, in contrast, focused on the environmental changes, political conflicts, and management challenges that emerge from flows of people, materials, and representations between cities and other spaces (P. Kirsten Valentine Cadieux and Hurley 2009; Kirsten Valentine Cadieux and Taylor 2013; Taylor & Hurley, 2016; Walker & Fortmann, 2003). These exurban studies have tended to frame conflicts in terms of the persistent differences between rural and urban identities, ways of life, and cultures and the diverse economies that underpin them, often using cultural landscape studies to focus on the construction of discourses and ideologies of nature. Both approaches have examined institutional power dynamics and the ways that politics shape decision-making processes, revealing a key element of political ecology approaches: Namely, a commitment to understanding drivers of social-ecological change and the environmental governance dynamics that emerge to "manage" this change.

Exurban political ecology and the politics of landscapes

Political ecology arrived in North America in the 1980s, with a focus on urban expansion – or the influence of migrants from cities – into traditionally rural areas and conflicts over land and resources, adopted from political ecology approaches to the developing world. For example, Fortmann (1996) called for using the tools of international property scholarship to help us understand conflicts over land and resources in the United States. Political ecologists focused on three themes or lenses in understanding urban expansion: ideologies of nature, production of protected places, and competing rural capitalisms.

Central to many of these political ecology studies of (ex)urbanization has been an examination of the attitudes and beliefs of individual landowners – in particular, their attitudes towards nature (J. Abrams and Bliss 2013; Kirsten Valentine Cadieux 2009; Hiner, 2014, 2016b; Johnson, 2008; Nesbitt and Weiner 2001). Urban expansion in many parts of the world has involved the arrival of new in-migrants, often from cities, in communities that for many decades had relatively stable, homogeneous populations (Hansen & Brown, 2005; Theobald, 2005). Political ecologists have hypothesized that these new arrivals bring with them new attitudes towards nature, which potentially shift how communities approach the regulation of land use and conservation (J. B. Abrams and Gosnell 2012; Beebe & Wheeler, 2012; Esparza & Carruthers, 2000; Hiner, 2015; Hurley & Walker, 2004). Yet some early research questioned the assumption that new arrivals are distinctively different in terms of their values and attitudes towards land use and land management (M. D. Smith and Krannich 2000), maintaining that attitudinal differences might more appropriately be attributed to economic marginality. Additionally, while political ecologists have tended to focus on the role of individual land owner attitudes, some have recently acknowledged that more attention should be paid to supply-side dynamics (i.e. the role of developers and the real estate industry) in the transformation of rural landscapes to exurban uses (Hurley, 2013) as well as the (sometimes unexpected) ways that actors engage in formal and informal regulatory and planning activities (Robbins, Martin, & Gilbertz, 2011).

Entwined with literature on exurban attitudes towards nature, a number of studies have examined the production of new protected places; e.g., political ecologists have pointed out paradoxes and how exurban migration to the urban fringe, often motivated by the desire for a greener lifestyle, necessarily changes the very landscapes that exurbanites seek (Kirsten Valentine Cadieux & Taylor, 2013). This means that exurban migrants, working together with some long-time locals, are often quick to advocate new conservation measures and seek to limit further growth and development in their newly adopted communities (Kirsten Valentine Cadieux 2009; Hurley & Walker, 2004; Johnson, 2008; Taylor & Hurley, 2016).

Another theme political ecologists use to conceptualize landscape transitions is competing rural capitalisms (P. Walker and Fortmann 2003). Characteristically exurban migrants move into landscapes that have been traditionally dominated by resource extractive industries such as mining, logging, and ranching. In many cases, these industries have experienced declines due to global rural restructuring (Woods, 2011). As a result, land values and opportunities for landowners to earn a living off the land are diminished. At the same time, accompanying the arrival of exurban migrants is often the rise of a new set rural industries tied directly to the visual consumption of natural amenities. Recreation, tourism, and rural real-estate development produce new landscape values, which can be conceptualized as a competing form of capital development, often viewed as incompatible with extractive industries (P. Kirsten Valentine Cadieux and Hurley 2009; Hiner, 2016b; Hurley & Arı, 2011; McKinnon, 2016; Taylor & Hurley, 2016; Walker & Fortmann, 2003).

Within the political ecology literature, there is also a largely unacknowledged body of work on the ways urbanization disrupts and transforms previously "rural" subsistence activities, including activities such as gathering non-timber forest products (Brown, 1995; Grabbatin, Hurley, & Halfacre, 2011; Hurley, Grabbatin, Goetcheus, & Halfacre, 2013; Hurley, Halfacre, Levine, & Burke, 2008; Robbins & Emery 2008). These works by political ecologists have documented the ways in which ecological and social changes associated with land ownership can create new hardships for rural peoples on the economic margins (Grabbatin et al., 2011; Hurley et al., 2008), and, contradictorily, how exurban property transformations sometimes open new opportunities for the persistence of natural resource livelihoods (Grabbatin et al., 2011; Hurley et al., 2013).

Today the main focus of political ecology in the developed world is exurbanization and amenity migration in the American West (Robbins et al. 2009). There are also an expanding number of case studies in other regions of the developed world, including other parts of the United States (Grabbatin et al., 2011; Hurley & Carr, 2010), Britain (Scott, Shorten, Owen, & Owen, 2009), Canada (Genevieve & Paradis, 2013; Guimond & Simard, 2010; Luka, 2013), and New Zealand (Kirsten Valentine Cadieux 2008). Robbins (2002) has suggested that political ecology needs to study up as well as down and to specifically examine the power of institutions and practices of officials while continuing a focus on what he calls the tools of political ecology, "ethnography and intense focus on micro-politics." Political ecology has tended to maintain this focus on ethnography and micro-politics while also examining the impacts of local dynamics on the politics of conservation. Yet this focus has perhaps obscured the need for work that examines the drivers of this global phenomenon and the social and environmental displacements these changes may cause (Gosnell, Abrams, and Abrams 2009; Newell & Cousins, 2014).

Urban political ecology and methodological cityism
As political ecology turned its attention to urban spaces and engaged directly with urban studies in the mid 2000s, it became reframed as urban political ecology. In an effort to engage with urban geography and re-nature urban processes, a seemingly new subfield or approach was created (Heynen, Kaika, and Swyngedouw 2006). Bringing the theoretical background and insights of political ecology to urban spaces, generally applied to places within the city, political ecologists have eagerly sought to dismantle the nature-culture divide by illuminating the myriad ways that cities are "natural"

(Gandy, 2002). A striking element of much early work in urban political ecology was its use of Marxist theory, particularly the concept of (urban) metabolism(s), to highlight the fundamental material links between country and city. However, as Heynen notes in his reviews of the development of UPE as a subfield (N. Heynen, 2014; N. Heynen 2015), these are not the only approaches now used, as feminist, post-colonial, post-humanist, and anti-racist approaches have challenged and enriched the field. (Gabriel, 2014; Gandy, 2012; Grove, 2009). It is also important to note that use of Marxist theory has been a part of political ecology from it start (Piers Blaikie & Brookfield, 1987; Peet & Watts, 1996)

In UPE, urban nature is theorized is through the concept of (urban) metabolism, which analyzes the flow of resources through the city and the mediations of such flows by economic, political, and social relationships (Cooke & Lewis, 2010). Newell and Cousins (2014) identify three separate lines of research that use the term "urban metabolism," including the one used in UPE. Urban metabolism, as used by urban political ecology, emerges directly from Marx's use of the term "Stoffwechsel" which literally translates from German as "change of matter" (Heynen et al., 2006). The term is used to describe the material processes that produce and reproduce urban spaces and ecologies. By drawing on this metaphor, urban political ecologists used the metabolism concept to trace the key ways and mechanisms through which urban space and its attendant biophysical dynamics were remade as cities grew through the constant turnover of land-uses. This approach helped to illuminate the ways that key actors (such as developers) and logics pervaded the creation of infrastructures and other elements needed to support city life while attending to the contradictions created by capitalism's need for ongoing growth and resources. Since most work in urban studies in the 20th century lacked a connection to ecological processes, early work in UPE focused on cities proper, often global cities, to lay the basic framework for how we can understand cities as both social and ecological creations. As Gandy (2015) points out, urban political ecology is only one of several lines of research into the historically contested character of urban ecologies.

Influenced by the work of David Harvey and Neil Smith, another central theme in many urban political ecology studies has been uneven development and the uneven production of green space. Uncovering the social and ecological processes that produce (access to) green space and other urban resources for some and not others fits well with UPE's focus on uncovering how nature is transformed in the city through social relations. UPE focuses on inequality in access to resources and spaces, taking on many of the same issues as environmental justice scholarship, but brings to the table a deep analysis of how the capitalist political-economic system is implicated in the production of such inequalities (N. Nik. et al., 2006). Heynen (2006) for example, analyzed changes in urban forest cover in Indianapolis between 1962 and 1993 using aerial photography to quantitatively measure the changes and Marxist political economy to provide analysis and explanation of shifting historical and political factors influencing landcover change in the city. Heynen (2006) found that changes in household income could be tied to shifts in residential forest cover. One of the strengths of UPE work has been to provide alternative explanations to liberal interpretations that tend to place blame for lack of green space on marginalized populations, demonstrating how broader political economies contribute to the production of greenspace (J. P. Evans, 2007; Hagerman, 2007; Nik Heynen, Perkins, & Roy, 2006; Quastel, 2009).

However, perhaps the greatest strength of UPE has been its focus on *power*, primarily understood as economic power (Domene, Saurí, and Parés 2005). Urban political ecology studies have examined how particular interests have been able to gain control of necessary resources and harness them, both materially and symbolically, for their own political projects, tying, for example, ecological processes to socio-political processes through commodification, privatization, and infrastructure building (Monstadt, 2009; Swyngedouw, 1997, 2004). Work from feminist, post-colonial, indigenous, and anti-racist perspectives in connection to UPE continues to enrich the approach, and points to areas of critical intersection with work in both urban and rural spaces on indigenous materialities (Larsen, 2016), food justice (K. Valentine Cadieux and Slocum 2015; K. Cadieux and Slocum 2015) and linked migration (Nelson & Nelson, 2010). We now turn our attention to the ways a unified urban political ecology and exurban political ecology strengthens our understanding of urbanization processes and how key questions of socio-economic processes produce particular forms of change and/or stability. We demonstrate how bringing insights from both these literatures together strengthens our understandings of urbanization and sustainability.

An integrated political ecology approach to urban rural interfaces

Urban political ecology has been separated from work on ex-urbanization largely through differing scalar and spatial foci (Figure 1). Yet some discernable methodological and theoretical tendencies can also be detected. Studies claiming the mantle of urban political ecology tend to take distinctly Marxist approach, focusing on cities to the exclusion of other urbanizing spaces and highlighting economic, political, and ecological processes taking place on the scale of urban regions. It is not uncommon now for scholars to call urban political ecology a paradigm, conceptual lens, sub-field, or approach (Cornea, Véron, & Zimmer, 2017; Holifield & Schuelke, 2015; Karpouzoglou & Zimmer, 2016; Silver, 2015), yet how this would differ from a "general" political ecology approach, is unclear. In contrast,

	Urban Political Ecology	Political Ecology of Exurbia
Spatial focus	Cities (i.e. methodological cityism)	Spaces in-between (cities, suburbs, exurban areas).
Methods	Analysis of social, political, and economic processes. Mix of archival, document analysis, interviews, and participatory methods.	The role of individual land owners, managers, and local politics. In-depth ethnographic case studies.
Scalar focus	City and megacity regions	Local case studies, imbedded in broader dynamics
Theory	Marxist metabolism, production of nature	Cultural landscape studies, representation of nature, competing rural capitalisms.
Themes	Environmental justice, uneven development, greenspace,	Impacts of urbanization, influence of urban on non-urban, production of protected places.
Strengths	Attention to power differences, justice, broad economic and political dynamics in development and conservation outcomes.	Attention to local knowledge, culture in development and conservation outcomes.

Figure 1. Comparing urban political ecology and exurban political ecology approaches.

literature on the ExPE has tended to borrow from cultural landscape studies, focusing on representations of nature and differing ideals at the local scale. In effect, the creation of UPE has, at least to some degree, reinforced the nature-society divide it was attempting to dissolve by reinforcing its analog, the urban-rural divide. Only a few studies in the urban political ecology tradition have worked across this spatial divide–or as some social-ecological scientists might suggest, this gradient–by focusing outside the city proper, particularly Robbins' (2003) work on lawns, Keil & Young's (2009) work on "in between" urban landscapes in Canada, and Swyngedouw (1999) and Kaika's (2005) work on the urbanization of water.

However, as some urban studies scholars have been influenced by the resurgent interest in Lefebvre's concept of a global urban society (see Brenner, 2013), there have been calls for UPE to give up its "methodological cityism" in favor of a new focus on urbanization processes (Angelo & Wachsmuth, 2015). Angelo & Wachsmuth (2015, p. 20) describe methodological cityism as "analytical privileging, isolation, and perhaps naturalization of the city in studies of urban processes where the non-city may also be significant." In this vein, it holds that, in order to be useful, the concept of planetary urbanization needs to pay attention to what is meant by "urban" and urbanization (R. Walker 2015) and acknowledge that "rural" is a category that continues to hold experiential and analytical power (Cloke 2006; Hiner, 2016c; Woods, 2011).

As part of the call to refocus on processes of urbanization beyond the city proper, a growing number of researchers have taken up work using a UPE framework to research sites outside of the city. Examples of this type of work include the work Gustafson (2015) on land use conflicts in exurban Appalachia, Kitchen's research on urban forests in South Wales (2013), and Parés, March, & Saurí (2013) study of the suburban landscapes of Barcelona. These new studies, which reach across this urban-exurban divide, have the potential to address the broader processes of globalization and uneven development. Gustafson et al. (2014) in particular proposes a new focus on megapolitan political ecologies, rightly highlighting the large ecological and social impacts of amenity migration, exurbanization, and rural gentrification outside of the urban center. However, in Gustafson's conceptualization of megapolitan political ecologies, urbanization processes are given explanatory power for changing land use and settlement patterns across a broad region. We contend that such UPE approaches would be further improved by a more explicit engagement with the existing literature in ExPE – counter to the trend toward *dis*engagement prevalent in the literature (McKinnon & Hiner, 2016).

Specifically, work on ExPE has the advantage of long standing engagement with places with a variety of relationships to urbanization and urbanism. First, ExPE, like its urban cousin, long has attended to issues of power and its influence on specific institutional decision-making arenas that shape the use of land and shape landscape change (Sandberg 2014; Hurley & Walker, 2004; P. A. Walker and Hurley 2011). Second, the ExPE also has continued political ecology's methodological focus on ethnography and micropolitics, focusing on the persistence of rural ecologies and livelihoods that get reworked, transformed, and conserved in highly uneven ways in particular places. We maintain that while it is necessary to examine the broad-scale ecological and social impacts of what has been called 'extended urbanization", it is not sufficient to stop there; rather it is key to examine how processes of extended urbanization, amenity migration, and rural gentrification *produce* uneven outcomes. These processes, occurring in places that are neither wholly rural nor urban, are particularly key because large areas of the Global North are being transformed into these low-density settlement and land use patterns. Once in place, exurban patterns appear to be highly resistant to further urbanization or densification as residents often maintain strong attachments to rural identities and regularly invoke strategies of conservation management (Taylor & Hurley, 2016).

In the section below, we present three case studies to demonstrate the contrasting and conflicting outcomes of the processes that shape exurbia – or extended urbanization – at the local level. We maintain that attention is needed to the specifics of local politics and ecologies if we are to understand whose vision of sustainability is being enacted in particular cases and what ecologies and cultures are being conserved. As the case studies below demonstrate, using insights and approaches from both lines of research strengthens studies in these contexts by bringing attention to processes occurring at multiple scales, effecting specific places, and uniting analysis that pay attention to material and cultural processes.

Three case studies of shifting dynamics on the urban-rural interface

Case 1: stone hill area exurbanites reproduce rural landscape aesthetics, mirroring former productive uses

In southeastern Pennsylvania, the "Stone Hill" area is an exurban enclave within the Philadelphia Metropolitan area. This area has experienced increasing residential development and urban migrants, yet the land use patterns of these new arrivals have tended to re-inscribe earlier forested and pastoral ecologies. Stone Hill is a ridgeline located in the western part of Montgomery County that stretches across several local townships. As a county identified "conservation landscape", a designation by the Montgomery County Planning Commission, Stone Hill has emerged as a conservation object where interventions by local municipalities are intended to protect open space through land acquisitions. Moreover, local municipalities have sought to maintain the area's rural characteristics through minimum lot-size zoning efforts. These efforts have been constrained by state court precedents (Hurley & Taylor, 2016), yet have contributed to the rise of an exurban pattern of residential development and associated advocacy efforts to formalize and expand the very conservation territories favored by county conservation landscape designations (Hurley, Maccaroni, & Williams, 2017). For example, urban in-migrants were instrumental in protecting nearly 100-acres of conserved open space through finding a conservation buyer, gaining commitments from two local townships for the purchase, and creating a public-private partnership to steward the forest. Besides social networking and political advocacy, one exurbanite has gone so far as to systematically purchase undeveloped lands for transfer into the conservation area. Yet these efforts also build on a history of expressed commitment to conservation by landowners with deep roots in the area, including landowners committed to rural recreational and natural resource uses. So in this case, rather than urban in-migrants conflicting with existing rural values, both groups have tended to work on conservation of traditional landscapes and uses, albeit unevenly.

Drawing on grounded visualization techniques (Hurley et al., 2008; Knigge & Cope, 2006), qualitative interviews, and air photo analysis, this research reveals the corresponding emergence of uneven land-use and management patterns on individual parcels associated with the exurban shift in the area. These patterns reveal the extent to which the exurban development process and household commitments hold divergent ideologies of nature that simultaneously reinscribe rural aesthetics into the area. In doing so, they point to the ways that flows of capital associated with urban in-migrants and the real estate markets they create produce uneven outcomes across this exurban landscape.

An ongoing and uneven shift from emphasis on natural resource uses to residential development has shaped land use and associated landscape changes in the area. Much of the area had either been converted to farmland, particularly in portions of the landscape outside of the ridgeline's characteristic boulder fields, or logged for various timber-related purposes by the end of the 19th Century. Yet, by the early 1940s, many of the areas of the ridgeline with extensive boulder fields had reforested and some smaller farms had been abandoned and begun reforesting. Beginning in the 1950s, early in-migrants to the area sought out historical homes associated with these small-scale farmsteads that had brought agriculture to the rocky slopes during the 18th Century. With the ability to commute by car to jobs in nearby towns, these individuals sought refuge from the expanding suburbs of eastern Montgomery County and access to lands to garden and harvest resources from the area's woodlands. In doing so, these individuals acted out early land-use practices that mirrored the rural practices of their neighbors at the time, including small-scale vegetable growing and some livestock tending. Moreover, these households maintained areas that would have otherwise returned to forest cover or reintroduced field openings to areas of the landscape that had recently reforested.

Continuing in-migration, however, eventually began to transform the landscape and more tightly link this exurban enclave to the city. Throughout the 1960s, 1970s, and 1980s, a small trickle of in-migrants arrived to build homes on smaller individual lots, introducing perforations into recovered forest through new openings for their homes and small yards. By the 1990s, larger parcels were becoming available for development, as various longstanding landowners passed away or decided to sell, and small, niche, large-lot subdivisions emerged in the area. In the process, parcels of cleared forest, semi-cleared forest, and fully forested areas became available for purchase to new in-migrants. Large-scale developers of traditional tract-style subdivisions had already leapfrogged the ridgeline for wider open and level land.

The increase in and shift toward a residential landscape introduced new ideas about land management to the area but in ways that continued to reflect past patterns of forest openings and pastoral aesthetics. By and large, households committed to forest stewardship have conserved and maintained areas of forest that have not been clear-cut since the end of the 19th Century (although selective harvests in these areas have changed these forests), including many areas of woodlands that were used during the early 20th Century for firewood harvest and sale to nearby urban centers. Meanwhile, households with suburban lawn commitments (see Robbins, 2007) have, together with specific developer interventions, generally maintained or reintroduced pastoral patterns of forest openings and aesthetics reminiscent of the smallholder farms that once characterized the area. These landowners espouse commitments to design features that maximize the amount of forest opening on their parcel for lawn and land management activities that prioritize suburban lawn aesthetics (Figure 2).

Instead of openings characterized by field crops or meadows, these non-forested areas are now maintained in turfgrasses and complimentary ornamental flower

Figure 2. Exurban Landscapes in Southeastern Pennsylvania. Back-to-the-Lander homestead (top) and cleared lawnscape bordering protected and forested open space. Lower Frederick Township, Pennsylvania.

plantings. Further, some residents in the area demonstrate landscape ideologies that prioritize explicit sustainability practices or intensive biodiversity conservation efforts. Those landowners committed to sustainability practices generally maintain parcels with pastoral land use patterns, signifying a continued commitment to forms of natural resource production that have been described as "back to the land" or "homesteading" dynamics in other areas. Meanwhile, landowners who espouse strong commitments to local forest types seek to maximize the amount of forest and native species gardening intended to create floral and faunal protection in line with their commitment to open space conservation efforts.

In this case of changing land use and management, on conservation lands and private residential parcels, the complexity of what are considered appropriate land management approaches by exurbanites becomes clear. These differences in approaches transcend categories of rural/urban landowners. Longtime residents from rural areas support new conservation and planning efforts, including efforts to construct the ridgeline into a conservation object worthy of recognition within county planning processes. Meanwhile, (predominantly) urban newcomers engage in the proliferation of urban vegetation dynamics, namely lawns, that recreate rural patterns of forest clearing and reimagine rural aesthetics in the process. So, too, both longtime rural residents and urban newcomers continue to turn to the land to find and extract natural resource values, including hunting, harvesting of non-timber forest products, and small-scale food production. Others incorporate classic rural animals, such as goats, into their lives as pets. These pets accompany their owners on hikes through the protected forest. Here the arrival of exurban migrants, rather than implementing a uniform pattern of urbanization, instead brings with it a complex of at least three different approaches to conservation and land management as well as a patchwork of settlement patterns, some of which reflect urban and suburban aesthetics. Meanwhile other landowners reenact longstanding rural livelihood and management strategies.

Case 2: emerging landscapes of wine production and consumption in the sierra nevada foothills

The foothills of the Sierra Nevada Mountains in California has a history of resource extraction, ranging from the infamous mid-1800s Gold Rush to timber production to cattle ranching (Duane, 2000; Momsen, 1996). In the past several decades, economic development has shifted towards housing development and tourism, although "heritage" uses continue to contribute significantly to the economy and local identities (Beebe & Wheeler, 2012; Hiner, 2014; P. Walker and Fortmann 2003). Meanwhile, a long-standing, but increasingly prominent hybrid use has been developing; wine grape production, an active agricultural use, paired with wine making, in conjunction with associated wine tourism, has been spreading in the landscape. The wine industry, "from grape to glass", is an increasingly visible, though largely un-quantified, economic force in the region. Ethnographic-style fieldwork was conducted in summer 2014 involving interviews with 60 wine grape growers, winemakers, winery owners, vineyard managers, agricultural advisors, wine retailers, and other people knowledgeable about the regional wine industry; participant observation; and a review of promotional and industry informational materials. This study of the area revealed that wine growers, wine makers, and wine buyers are engaged in an exchange that links rural and urban together in a mutual – if perhaps uneven – economic and cultural interchange, such that urban consumers and investors are set to gain more than their rural counterparts. However, that said, rural actors also actively engage in the changes taking place by producing a both wine and wine landscapes for the consumption of urban markets.

While it is largely urban tourists who visit the area for wine-related activities (and, indeed, new vineyard owners and wine makers are often urban transplants as well), wine-based activities are closely tied to the rural appeal of the place, capitalizing on the rolling hills, oak woodlands, and cattle-strewn landscape to both draw in visitors and in-migrants (Figure 3). The success of an "emerging" wine region such as the Sierra Nevada foothills rests not just on the abilities of wine growers and wine makers to produce a quality product, but also on their ability to successfully market it, namely by luring urban consumers with the aesthetics of the wine landscape (vineyards, wineries, tasting rooms, etc.) as well as their place-soaked product. Direct sales are, of course, only part of the marketing strategies of many producers (producers who may also distribute their

Figure 3. Exurban Sierra Nevada foothill landscape with rolling hills dotted with trees, cattle, and, increasingly, vineyards. Calaveras County, California.

product at a variety of scales through stores, restaurants, and other markets), but direct sales to on-site consumers amounts to a significant portion of their appeal.

Wine is a product that reflects and produces local ecologies and environments in very specific ways (Dougherty 2012). Terroir is an essential component of wine growing and making; the (micro)climactic, geologic, and environmental characteristics of a place are intimately tied to which kinds of varietals can be produced where and at which quality (Trubek & Bowen, 2008; Unwin, 2012). Yet as landowners, land managers, and investors pursue wine as a land use and economic strategy, certain activities and actors are preferred over others, producing new ecologies and environments. Uses that may be long-standing but are no longer reliably profitable may be sidelined. Ranches or orchards turn to vineyards, barns turn to wineries and tasting rooms, and hillsides become caves or cellars (Hiner, 2016a).

Moreover, wine tasting is an exercise in embodied place consumption. Wine enthusiasts visit vineyards and wineries to consume not just wine, but also the visual and aesthetic properties of the place. Wine tourists drink in the landscape as they travel from tasting room to tasting room and consume the product of that place directly through the wine. Wine sold in tasting rooms is not always locally-sourced, and the share of wine that is made from non-local grapes varies by region, but, in the Sierra Nevada, most of the wine grapes used to produce "Sierra" wine is (still) sourced from within the region. The ability of wine retailers to do well in their business is directly related to consumer experiences and perceptions of the place they are visiting. As such, maintaining (and building) such a place-to-be-consumed is an ongoing social process. In other words, the Sierra Nevada is a productive, active landscape, but it is also one which is (re)created for the pleasure of incoming visitors and/or migrants – even sometimes at the expense or displeasure of previous residents or stakeholders.

While the production of a wine landscape is undoubtedly a transformative process environmentally, economically, and socially, the industry also protects and produces coveted – and often idealized and/or imagined – rural landscapes and values. The influx of and deference to wine consumers encourages a certain kind of rural placemaking, wherein the functional, productive landscape is leveraged to build idealized landscapes that cater to urban environmental imaginaries. The process whereby commodities are produced in rural places and are then sold and distributed to urban ones is altered such that urban consumers are further privileged; urbanites move beyond billing externalities to the rural communities that sustain them into an exchange whereby they consume not only rural products but rural place itself (Hurley, 2013).

In this way, the Sierra wine region is a place where urban desires and imaginaries increasingly dominate, as (agri)cultural products are transformed into financial and symbolic capital (Hiner, 2016a; Sayre, 2002) and the idyllic/idealized rural landscape is commodified. Urban and rural, as such, are tied together and artificially separating them hinders rather than helps analysis of the processes occurring there.

In sum, while the urban is increasingly "everywhere", we maintain that in rural areas along the urbanizing fringe rural imaginaries remain significant for cultural (re)production, political negotiations, and environmental management decisions – especially in those areas feeling pressure from proximate urban zones. The insights of ExPE related to contested politics and environments including discussions of competing rural capitalisms, the preservation and creation of conservation landscapes, and the ideological and material power of rural idylls, as well as the insights from urban political ecology related to power, privilege, and the metabolism of nature are both useful here. Emerging wine landscapes like those in the Sierra Nevada provide insights into the social, economic, and environmental processes that tie cities to other spaces in ever more complex ways.

Case 3: urbanization without urbanism: uneven urban metabolisms in jackson county oregon

Jackson County, located in southern Oregon, is a small metropolitan area with a polycentric, sprawling development pattern. While Jackson County hardly constitutes an urban area in the minds of most Oregonians, this small metropolitan region is part of what Luke (2003) calls "global cities", where most of the world's urban population still lives. Population growth in Jackson County depends on a flow of migrants from large urban areas into the small cities and rural areas in the county. Yet growth in Jackson County cannot be conceptualized simply as a matter of counter-urbanization or de-urbanization. Neither can it be understood as a straightforward embrace of urbanization.

Neither the numbers nor the urban origins of migrants fully captures the role of the rural in promoting urbanization. What emerges from both written documentation of land use planning processes and interviews with local residents is how new arrivals value this place for its *rural characteristics* and desire the preservation of those qualities. It is this attachment to "ruralism" and rejection of urban values that limits how growth takes place and promotes policies that contradict traditional visions of urban sustainability such as density of urban form, transit oriented and mixed use development, and separation of urban and rural uses.

Traditionally, the economy of Jackson County relied on constantly varying levels of mining, forestry in the surrounding mountains, and pear growing on the valley floor along with longstanding low levels of tourism and rural residential development (McKinnon, 2016). In the early 20th century hundreds of small orchard growers filled the valley with fruit trees, making the region one of the largest pear producing districts in North America. The region was filled with a fervor for the Jeffersonian ideal based on small farms, but starting in the 1930s farmland gradually consolidated into the hands of a few large growers, so that by the time of this study, only a handful of independent pear growers remained. Large growers made their money largely from packing and shipping, rather than growing. By the 1980s the few remaining small growers faced increased competition from fruit growers in Asia and Latin America leading to stagnant prices and increasing conflict with urban and exurban neighbors.

By the early 2000s significant patches of land in the valley continue to be farmed, but the expanding urban footprint of Medford in particular, has swallowed up significant portions of the rural landscape. Yet rapid urbanization in California beginning post-WWII fueled a distinctly rural and dispersed development pattern in Jackson County as back-to-the-landers and white flight increased the significance of the rural idyll – and, accordingly, the blossom-filled valley, surrounded by deep green slopes, appealed to increasing numbers of new arrivals. By the 1990s, these new arrivals along with many of the region's remaining farmers begin a grassroots planning process out of concern over sprawling urbanization in the valley. Yet over the next two decades, as they worked through development of a regional conservation plan, they were largely unable to escape increasingly urbanized patterns of development. Large scale, master planned developments now predominate and, while there are efforts towards sustainability through higher density, new urbanism, and transit connections, these urban planning strategies require large parcels of farmland. Oregon's statewide planning regime mandates that urban expansion take in low density sprawl and preserve areas of intact farm and forest landcover. However, in actuality the types of large scale residential, commercial, or retail development in demand in the growing region would be prohibitively expensive if developers attempted to purchase the many small parcels required from individual land owners. In part due to the need to redevelop the physical infrastructure supporting water, sewage, and power to these exurban enclaves. Additionally, exurban residents resist any attempt to annex their lands by cities in the region whereas farmers, most of whom are now aging out of the profession, are often eager to sell.

In 2013, six municipalities in the region adopted the Greater Bear Creek Regional Problem Solving plan, which established urban reserves, designating lands for urban growth over the next 50 years, and successfully concluding over 20 years of collaborative plan development. The plan halts further urbanization of the most fertile farmlands in the valley, the areas along the riparian floodplain of the Bear Creek, which flows north from Ashland through the largest community, Medford and into the Rogue River in the northern portion of the valley. Yet this land is largely already covered in an exurban residential development pattern, with remaining commercial farm plots gradually giving way to a mix of residential uses and post-productivist agriculture (N. Evans, Morris, and Winter 2002; Holmes, 2002), which relies of the proximity of urban consumers even as it trades on the desire for rural experiences.

New arrivals engage in small-scale production on their properties but often with a focus on the experience of farming or rural life instead of commodity production (Cadieux, 2008). The growing number of new arrivals with urban tastes for wine and specialty gourmet foods in Medford and Ashland opens up new markets for specialized agricultural production and the consumption of rural experiences. This can be seen in the growth of direct marketing, farmers' markets, local food production, vineyards and wineries that provide food and wine for consumers willing to pay for not only the product but also for the experience of visiting the farm or the farmers market (see Figure 4). In this way, pear farms are being replaced by a mix of luxury equestrian ranches, small vineyards, and suburban and exurban homesteaders keeping their own chickens.

This rise in new "hobby farms" produces a secondary transformation as farm suppliers and tractor dealers have been replaced by a growing secondary industry providing supplies and assistance designed specifically for recreational farming. For example, the many micro-vineyards in the region are serviced by vineyard management companies, which allow would-be winemakers to enjoy the dream of living on a rural estate with its own vineyard while the work of growing the grapes and making the wine is taken care of by others. The finished wine, bottled and labeled, is brought back to the owners for sale or private distribution to friends and family. This trend toward postage-stamp wineries is mirrored in other emerging wine regions around the United States and beyond.

Yet for all their professed desire to escape urban life, exurbanites continue to demand urban levels of social provisioning and consumption. Medford has become the center for retail and medical services for an expansive rural region. The growing urban desires and tastes of the population can also be seen in the increasing sophistication and urban orientation of consumption in the valley, for example, the arrival of REI in the valley in 2012 (see Figure 5).

The combination of increasingly urbanized metabolic processes in the economy and the marketing of rural idylls for urban consumption has created an urban form that is

Figure 4. Marketing rural space, Hillcrest Winery, Medford Oregon.

Figure 5. New mall under construction in west Medford, anchored by REI and Trader Joes.

sprawling and a local political climate that resists attempts to impose urban planning solutions such as increased density, transit oriented development, and the separation of urban and rural uses. While understanding the power of urban capital and urban metabolisms in this situation is key, it is not sufficient to fully explain development patterns. Rural idylls continue to shape regional development and patterns of urbanization.

Urban sustainability in an urbanizing world

Political ecology offers insights into how power functions to enable or constrain particular processes and outcomes. Understanding such economic and social processes is key to discussions of sustainability. Yet the focus on a limited subset of urban forms and processes within sustainability discussions limits our ability to understand how processes of urbanization produce uneven impacts across the landscape, gentrification in one location and concentrated poverty in another, green spaces for some and environmental degradation for others.

However, a divided literature within political ecology tends to limit its usefulness for issues of urban sustainability. This division is, in some ways, to be expected; uneven development produces a world in which privilege and deprivation are often strongly spatially differentiated. Additionally, political ecology studies tend to be strongly tied to places and processes at the local scale, reflecting the fields strong reliance on case studies and commitment to grounding theory in particular locations.

A multiscalar focus on the broader processes at work and how these intersect within particular places to produce the uneven outcomes is a key strength of political ecology approaches. While some urban political ecologists have taken an important step forward in moving away from methodological cityism, additional steps are needed to further develop a united political ecology of (ex)urbanization. In such a sprawling field, segregating research foci by geographic location or resource type may be seen both as easy and appropriate, but dialog and engagement across the divide is essential. This will mean that as UPE moves away from a focus on cities and towards a focus on urbanization processes, it will need to engage already existing bodies of literature on non-urban, ex-urban, and *zwischenstadt* landscapes (Sieverts, 2003). These literatures include significant work by political ecologists on the cultural politics of amenity migration (Cadieux & Taylor, 2013; Walker & Fortmann, 2003) and exurbanization (Taylor & Hurley, 2016). Additionally, as research on the persistence of rural activities in rapidly urbanizing areas reminds us, it is not only the symbolic dimensions of rural idyll aesthetics at play. Scholars of sustainability also need to pay better attention to the ways that changes created by urbanization continue to incorporate existing or enable new productive dimensions of natural resource use, including among economically marginalized groups (Grabbatin et al., 2011; Hurley & Taylor, 2016).

To move away from an exclusive focus on cities and the concrete and clay dimensions of the built environment, UPE must better understand the ways rural ideals and ideologies of nature both continue to shape and reshape urbanization processes – particularly as urbanization processes increasingly extend beyond what are widely recognized as urban landscapes. Moreover, as traditionally rural activities increasingly move into the city (Cantor, n.d.; Cloke, 2006; Lacour & Puissant, 2007) they become a focus for urban sustainability research. These activities include urban agriculture (Colasanti, Hamm, & Litjens, 2012) and foraging in urban green-spaces (R. J. McLain et al. 2013; R. McLain et al. 2012; Poe, LeCompte, McLain, & Hurley, 2014; Poe, McLain, Emery, & Hurley, 2013).

At the same time, researchers steeped in the literature on exurban and rural resource conflicts would benefit from theoretical engagements with global urbanization. In particular, engagement with the literature in urban political ecology would shift the focus from the discourses used by exurbanites and locals by situating those discourses within flows of capital and materials. Abrams and Gosnell (2012) have suggested that while we now know a significant amount about amenity migrants themselves, we know less about the other actors involved in facilitating the "green sprawl" process such as real estate developers, local boosters, builders, landowners, and speculators.

Expanding conceptions of urban sustainability to 1) contemplate the broad range of settlement types that are being created across landscape gradients as part of urbanization and 2) examine how patterns in one place may or may not be related to patterns in another place (and the flows of people, ideas, and capital in-between) would be of advantage to both researchers and activists alike. Research on sustainability cannot afford to focus solely on the urban core because measuring sustainability at one scale and location potentially misses the displacement of other impacts. Further, continuing to focus on a simplistic urban-rural dichotomy or focusing only on sustainability within the urban core constrains our capacity to consider potential solutions for resource-intensive land uses. Moreover, the creation of an integrated political ecology of (ex) urbanization would facilitate an increased understanding of socio-ecological processes and management approaches across scales, returning to the strengths of early political ecology studies (Robbins and Monroe Bishop 2008).

The place of political ecology in urban sustainability

Geographers long have examined human-environment interactions and their consequences for society, drawing on various schools of thought and theoretical framings

(Harden, 2012; Turner & Robbins, 2008). Competing framings have considered the effect of humans on nature and the effect of nature on humans in various ways. But as Harden (2012, p. 742) notes, the overwhelming philosophical approach within geography has been one where humans as are seen as "separate from nature" (see also Heynen, Kaika, & Swyngedouw, 2006; Smith, 2008). As a result, the seemingly obvious dichotomy has served to obscure the actual complexity of interactions and feedbacks between humans and nonhumans. In contrast, new theoretical approaches emerging within human-environment geography and allied fields seek to integrate the study of humans and nature (Turner & Robbins, 2008). Land-change science analyses of the drivers of environmental change and their effects on Earth's systems and ecosystem service provisions (Turner & Robbins, 2008). Social-ecological systems' focuses on coupled human natural systems and their resilience to perturbations (Cumming, 2011). A common theme among these frameworks is exploration of the role that social and ecological dynamics play in creating bidirectional effects (Turner & Robbins, 2008). A key analytical advantage of these approaches to sustainability is understanding the ways that natural limits shape social responses, the role of complexity, and produce emergent responses in both human and natural systems adapting to changing conditions (Cumming, 2011; Turner & Robbins, 2008). Yet political ecologists (and other social scientists) have suggested these studies are insufficient to fully understand the complex ways that human institutions and human-environment interventions shape sustainability practices (Cumming, 2011 Isenhour, McDonogh, & Checker, 2015; Turner & Robbins, 2008). Rather they have sought to document the grounded human practices that create sustainable places and land uses.

Land-change science and other positivist approaches, including those of urban social ecological systems scholars, broadly construed, go a long way to addressing key questions of urban sustainability (Elmqvist et al., 2013; Turner & Robbins, 2008). At the same time, in balancing the ecological with the social, political ecologists repeatedly have insisted that these approaches may miss key insights about the social factors that either enable or constrain actors within diverse institutional or decision-making contexts at various scales and their ability to draw on different degrees of political and economic power (Taylor & Hurley, 2016; Turner & Robbins, 2008). Still, as Turner and Robbins (2008, p. 300), speaking specifically about the relationship between lands change science and political ecology, suggest: both land-change science and political ecology "follow land management practices to their environmental consequences, although each expresses this concern differently."

While acknowledging the critical work of land change science and other positivist approaches, we have endeavoured to demonstrate the ways that an integrated political ecology further illuminates the key social dynamics shaping (un)sustainable land change. In so doing, we center our efforts on the ways that (formerly) rural places are transformed by the interrelated dynamics of capital flows and ideological interpretations of material nature that accompany the movement of people to these spaces (see also Taylor & Hurley, 2016.) Political ecology helps us to better understand not only flows of material and energy but also who – or what environmental imaginaries – control nearby spaces, corresponding (de)legitimated land uses, and associated products.

Through this review and examination of case studies, we hope to encourage renewed engagement between political ecologists with differing locational and theoretical commitments and increased collaboration between political ecologists and other researchers working on urban sustainability issues (Turner & Robbins, 2008). We acknowledge that there are many challenges to engaging across various theoretical and political differences (Blaikie, 2012); however, our understandings of urban sustainability can only be deepened through cross disciplinary conversation and collaboration.

Disclosure statement

No potential conflict of interest was reported by the authors.

Funding

Support for field research in southeastern PA was provided by the Office of the Dean of Academic Affairs at Ursinus College. Funding for field research in the Sierra Nevada Mountains in California was provided by the Research Enhancement Program at Texas State University. Field work in Southern Oregon was supported by the Larry Ford Fieldwork Scholarship for Cultural Geography.

ORCID

Innisfree McKinnon ⓘ http://orcid.org/0000-0003-4670-5674
Colleen C Myles ⓘ http://orcid.org/0000-0002-1181-8911

References

Abrams, Jesse B., & Gosnell, Hannah. (2012). The politics of marginality in Wallowa County, Oregon: Contesting the production of landscapes of consumption. *Journal of Rural Studies, 28* (1), 30–37. Elsevier Ltd.
Abrams, Jesse, & Bliss, John C. (2013). Amenity landownership, land use change, and the re-creation of 'working landscapes. *Society & Natural Resources, 26*(7), 845–859. Routledge.
Anders, Sandberg, L. (2014, January). Environmental gentrification in a post-industrial landscape: The case of the Limhamn Quarry, Malmö, Sweden, *Local Environment*, Routledge,19 (10) 1–18.
Angelo, Hillary, & Wachsmuth, David. (2015). Urbanizing urban political ecology: A critique of methodological cityism. *International Journal of Urban and Regional Research, 39*(1), 16–27.
Beebe, Craig, & Wheeler, Stephen M. (2012). Gold country : The politics of landscape in Exurban El Dorado County, California. *Journal of Political Ecology, 19*, 1–16.
Blaikie, Piers. (2008). Epilogue: Towards a future for political ecology that works. *Geoforum, 39* (2), 765–772.
Blaikie, Piers. (2012). Should some political ecology be useful? The inaugural lecture for the cultural and political ecology specialty group, annual meeting of the association of American geographers, April 2010. *Geoforum, 43*(2), 231–239.
Blaikie, Piers, & Brookfield, Harold. (1987). *Land degradation and society*. Routledge Kegan & Paul, Methuen & Co. Ltd, NY, NY.
Brenner, Neil. (2013). Theses on Urbanization. *Public Culture, 25*(1 69), 85–114.
Brenner, Neil, & Schmid, Christian. (2014). The 'urban Age' in Question. *International Journal of Urban and Regional Research, 38*(3), 731–755.
Brown, Beverly. (1995). *In timber country*. Temple University Press, Philadelphia.
Cadieux, Kirsten Valentine. (2008). Political ecology of Exurban 'lifestyle' landscapes at Christchurch's contested urban fence. *Urban Forestry & Urban Greening, 7*(3), 183–194.

Cadieux, Kirsten Valentine. (2009). Competing discourses of nature in Exurbia. *GeoJournal, 76 (4), 341-363.*
Cadieux, Kirsten Valentine, & Hurley, Patrick T. (2009). Amenity migration, Exurbia, and emerging rural landscapes: Global natural amenity as place and as process. *GeoJournal, 76 (4), 297-302.*
Cadieux, Kirsten Valentine, & Slocum, Rachel. (2015a). What does it mean to do food justice? *Journal of Political Ecology, 22, 1-26.*
Cadieux, Kirsten Valentine, & Taylor, Laura. (2013). *Landscape and the ideology of nature in Exurbia: Green Sprawl.* New York, NY: Routledge.
Cadieux, Kirsten. Valentine, & Slocum, Rachel. (2015b). Notes on the practice of food justice in the U.S.: Understanding and confronting Trauma and inequity. *Journal of Political Ecology,22, 27-52.*
Cantor, Alida. (2017). Urbanization beyond the city: A relational urban political ecology of Southern California's hydrosocial hinterlands.
Cloke, Paul J. (2006). Conceptualizing Rurality. In Paul J. Cloke, Terry. Marsden, & Patrick H. Mooney (Eds.), *Handbook of rural studies* (pp. 18-28). SAGE, Thousand Oaks CA.
Colasanti, Kathryn, Hamm, Michael W., & Litjens, Charlotte M. (2012). The city as an 'Agricultural Powerhouse'? Perspectives on expanding urban agriculture from Detroit, Michigan. *Urban Geography, 33*(3), 348-369.
Cooke, Jason, & Lewis, Robert. (2010). The nature of circulation: The urban political ecology of Chicago's Michigan Avenue Bridge, 1909-1930. *Urban Geography, 31*(3), 348-368.
Cornea, Natasha, Véron, René, & Zimmer, Anna. (2017). Clean city politics: An urban political ecology of solid waste in West Bengal, India. *Environment and Planning A* 49(4), 728-744. .
Cumming, Graeme S. (2011). Spatial resilience: Integrating landscape ecology, resilience, and sustainability. *Landscape Ecology, 26*(7), 899-909.
Dirksmeier, Peter. (2012). The wish to live in areas with 'people like Us': Metropolitan habitus, habitual urbanity and the visibility of urban-rural differences in South Bavaria, Germany. *Visual Studies, 27*(1), 76-89. Routledge.
Domene, Elena, Saurí, David, & Marc, Parés. (2005). Urbanization and sustainable resource use: The case of garden watering in the metropolitan region of Barcelona. *Urban Geography, 26*(6), 520-535.
Dougherty, Percy H. (2012). *The geography of wine: regions,terroir and techniques.* Dordrecht: Springer Netherlands.
Duane, Timothy P. (2000). *Shaping the Sierra: Nature, culture, and conflict in the changing West* (1st ed.). University of California Press, Berkeley and Los Angeles CA.
Elmqvist, Thomas, Michail Fragkias, Julie Goodness, Burak Güneralp, Peter J. Marcotullio, Robert I. McDonald, Susan Parnell, Maria Schewenius, Marte Sendstad, Karen C. Seto, Cathy Wilkinson(Eds.), *Urbanization, biodiversity and ecosystem services: Challenges and opportunities.* Dordrecht: Springer Netherlands.
Esparza, Adrian X., & Carruthers, John I. (2000). Land use planning and exurbanization in the rural mountain West: Evidence from Arizona. *Journal of Planning Education and Research, 20* (1), 23-36.
Evans, James P. (2007). Wildlife corridors: An urban political ecology. *Local Environment, 12*(2), 129-152. Routledge.
Evans, Nick, Morris, Carrol, & Winter, Michael. (2002). Conceptualizing agriculture: A critique of post-productivism as the New Orthodoxy. *Progress in Human Geography, 26*(3), 313-332.
Fortmann, Louise. (1996). Bonanza! The unasked questions: Domestic land tenure through international lenses. *Society & Natural Resources, 9*(5), 537-547.
Furuseth, Owen J., & Lapping, Mark B. (1999). *Contested countryside: The rural urban fringe in North America.* United Kingdom: Ashgate.
Gabriel, Nate. (2014). Urban political ecology: Environmental imaginary, governance, and the non-human. *Geography Compass, 8*(1), 38-48.
Gandy, Matthew. (2002). *Concrete and clay: Reworking nature in New York City* (1st ed.). Cambridge, MA: The MIT Press.

Gandy, Matthew. (2012). Queer ecology: Nature, sexuality, and heterotopic alliances. *Environment and Planning D: Society and Space, 30*(4), 727–747. SAGE Publications.

Gandy, Matthew. (2015). From Urban Ecology to Ecological Urbanism: An Ambiguous Trajectory. *Area, 47*(2), 150–154.

Genevieve, Vachon, & Paradis, David. (2013). Design and conservation in quebec city's rural-urban fringe: The case of lac-beauport. In Kirsten Valentine Cadieux & Laura Taylor (Eds.), *Landscape and the ideology of nature in Exurbia: Green Sprawl* (pp. 159–184). Abingdon-on-Thames, UK: Routledge.

Gosnell, Hannah, Abrams, Jesse, & Jesse Abrams, Æ. (2009). Amenity migration: Diverse conceptualizations of drivers, socioeconomic dimensions, and emerging challenges. *GeoJournal, 76*(4), 1–20. Springer Netherlands.

Gottmann, Jean. (1957). Megalopolis or the urbanization of the Northeastern Seasboard. *Economic Geography, 33*(3), 189–200.

Grabbatin, Brian, Hurley, Patrick T., & Halfacre, Angela. (2011, July). 'I still have the old tradition': The co-production of sweetgrass basketry and coastal development. *Geoforum, 42*(6), 638–649. Elsevier Ltd.

Grove, Kevin. (2009). Rethinking the nature of urban environmental politics: Security, subjectivity, and the non-human. *Geoforum, 40*(2), 207–216.

Guimond, Laurie, & Simard, Myriam. (2010). Gentrification and neo-rural populations in the québec countryside: Representations of various actors. *Journal of Rural Studies, 26*(4), 449–464.

Gustafson, Seth. (2015). Maps and contradictions: Urban political ecology and cartographic expertise in Southern Appalachia. *Geoforum, 60*, 143–152.

Gustafson, Seth, Nik Heynen, Jennifer L., Rice, Ted Gragson, Marshall Shepherd, J., & Strother, Christopher. (2014). Megapolitan political ecology and urban metabolism in Southern Appalachia. *The Professional Geographer, 66*(4), 664–675. Routledge.

Hagerman, Chris. (2007). Shaping neighborhoods and nature: Urban political ecologies of urban waterfront transformations in Portland, Oregon. *Cities, 24*(4), 285–297.

Hansen, Andrew J., & Brown, Daniel G. (2005). Land-use change in rural America: Rates, Drivers, and consequences. *Ecological Applications, 15*(6), 1849–1850.

Harden, Carol P. (2012, June). Framing and reframing questions of human–environment interactions. *Annals of the Association of American Geographers, 102*(4), 737–747. Taylor & Francis Group.

Heynen, Nick. (2014). Urban political ecology I: The urban century. *Progress in Human Geography, 38*(4), 598–604.

Heynen, Nick. (2015, November). Urban political ecology II: The abolitionist century, *Progress in Human Geography, 40*(6), SAGE Publications, 309132515617394.

Heynen, Nik. (2006). Green urban political ecologies: Toward a better understanding of inner-city environmental change. *Environment and Planning A, 38*(3), 499–516.

Heynen, Nik, Perkins, Harold A., & Roy, Parama. (2006). The political ecology of uneven urban green space. *Urban Affairs Review, 42*(1), 3–25.

Hiner, Colleen C. (2014). 'Been-Heres vs. Come-Heres' and other identities and ideologies along the rural–urban interface: A comparative case study in Calaveras County, California. *Land Use Policy, 41*(November), 70–83.

Hiner, Colleen C. (2015). (False) Dichotomies, political ideologies, and preferences for environmental management along the rural-urban interface in Calaveras County, California. *Applied Geography, 65*, 13–27.

Hiner, Colleen C. (2016a). 'Chicken Wars,' water fights, and other contested ecologies along the rural-urban interface in California's Sierra Nevada Foothills. *Journal of Political Ecology, 23*, 167–181.

Hiner, Colleen C. (2016b). Divergent perspectives and contested ecologies: Three cases of land-use change in Calaveras County, California. In Laura E. Taylor, Patrick T. Hurley (Eds.), *A comparative political ecology of Exurbia* (pp. 51–82). Cham: Springer International Publishing.

Hiner, Colleen C. (2016c). Beyond the edge and in between: (Re)conceptualizing the rural–urban interface as meaning–model–metaphor. *The Professional Geographer, 68*(4), 520–532. Routledge.

Holifield, Ryan, & Schuelke, Nick. (2015). The place and time of the political in urban political ecology: Contested imaginations of a river's future. *Annals of the Association of American Geographers, 105*(2), 294-303.

Holmes, John. (2002). Diversity and change in Australia's rangelands: A post-productivist transition with a difference? *Transactions of the Institute of British Geographers, 27*(3), 362-384.

Hurley, Patrick T., & Arı, Yılmaz. (2011). Mining (Dis)amenity: The political ecology of mining opposition in the Kaz (Ida) mountain region of Western Turkey. *Development and Change, 42*(6), 1393-1415.

Hurley, Patrick T, & Carr, Edward R. (2010). Introduction: Why a political ecology of the U.S. South? *Southeastern Geographer, 50*(1), 99-109.

Hurley, Patrick T., Grabbatin, Brian, Goetcheus, Cari, & Halfacre, Angela. (2013). Gathering, buying, and growing sweetgrass (Muhlenbergia Sericea): Urbanization and social networking in the sweetgrass basket-making industry of lowcountry South Carolina. In Robert Voeks, John Rashford (Eds), *African ethnobotany in the Americas* (pp. 153-173). New York, NY: Springer New York.

Hurley, Patrick T., Halfacre, Angela C., Levine, Norm S., & Burke, Marianne K. (2008). Finding a 'Disappearing' nontimber forest resource: Using grounded visualization to explore urbanization impacts on sweetgrass basketmaking in greater Mt. Pleasant, South Carolina. *Professional Geographer, 60*(4), 556-578.

Hurley, Patrick T., Maccaroni, Megan, & Williams, Andrew. (2017). Resistant actors, resistant landscapes? A historical political ecology of a forested conservation object in Exurban Southeastern Pennsylvania. *Landscape Research, 42*(3), 291-306. Routledge.

Hurley, Patrick T., & Taylor, Laura E. (2016). Conclusion: Moving beyond competing rural capitalisms and uneven environment management in Exurbia. In Laura Taylor, Patrick T Hurley (Eds), *A comparative political ecology of Exurbia* (pp. 285-301). Cham: Springer International Publishing.

Hurley, Patrick T., & Walker, Peter A. (2004). Whose vision? Conspiracy theory and land-use planning in Nevada County, California. *Environment and Planning - Part A, 36*(9), 1529-1547.

Hurley, Patrick T. (2013). Whose sense of place? A political ecology of amenity development. In William P. Stewart, Daniel Williams, & Linda Kruger (Eds.), *Place-based conservation: Perspectives from the social sciences* (pp. 165-180). New York, NY: Springer.

Isenhour, Cindy, McDonogh, Gary, & Checker, Melissa. (2015). *Sustainability in the global city*. Cambridge, MA: Cambridge University Press.

Jansson, André. (2013). The hegemony of the urban/rural divide: Cultural transformations and mediatized moral geographies in Sweden. *Space and Culture, 16*(1), 88-103.

Johnson, Brian Edward. (2008). Nature, affordability, and privacy as motivations for Exurban living. *Urban Geography, 29*(7), 705-723.

Kaika, Maria. (2005). *City of flows: modernity, nature, and the city* (1st ed.). Abingdon-on-Thames, UK: Routledge.

Karpouzoglou, Timothy, & Zimmer, Anna. (2016). Ways of knowing the wastewaterscape: Urban political ecology and the politics of wastewater in Delhi, India. *Habitat International, 54*(May), 150-160.

Keil, Roger. (2005). Progress report—urban political ecology. *Urban Geography, 26*(7), 640-651. Bellwether Publishing.

Keil, Roger, & Young, Douglas. (2009, August). Fringe explosions: Risk and vulnerability in Canada's new in-between urban landscape. *Canadian Geographer/Le Géographe, 53*(4), 488-499. Canadien.

Kitchen, L. (2013). Are trees always "good"? urban political ecology and environmental justice in the valleys of south wales. *International Journal of Urban and Regional Research, 37*(6),1968-1983. http://doi.org/10.1111/j.1468-2427.2012.01138.x doi:10.1111/j.1468-2427.2012.01138.x

Knigge, LaDona, & Cope, Meghan. (2006). Grounded visualization: Integrating the analysis of qualitative and quantitative data through grounded theory and visualization. *Environment and Planning A, 38*(11), 2021–2037.

Krause, Monika. (2013). The ruralization of the world. *Public Culture, 25*(2 70), 233–248.

Lacour, Claude, & Puissant, Sylvette. (2007). Re-urbanity: Urbanising the rural and ruralising the urban. *Environment and Planning, 39*(1), 728–748.

Larsen, Soren C. (2016). Regions of care: A political ecology of reciprocal materialities. *Journal of Political Ecology, 23*, 159–166.

Luka, Nik. (2013). Sojourning in nature: The second-home exurban lanscapes of Ontario's Near North. In Kirsten Valentine Cadieux & Laura Taylor (Eds.), *Landscape and the ideology of nature in Exurbia: Green Sprawl*, New York, NY: Routledge (pp. 121–158).

Luke, Timothy W. (2003). Global cities Vs.' global cities:' Rethinking contemporary urbanism as public ecology. *Studies in Political Economy, 70*(1), 11–33. Retrieved from https://www.mediatropes.com/index.php/spe/article/view/12074

McKinnon, Innisfree. (2016). Competing or compatible capitalisms? Exurban Sprawl and high-value agriculture in Southwestern Oregon. In Laura E. Taylor & Patrick T. Hurley (Eds.), *A comparative political ecology of exurbia: Planning, environmental management, and landscape change*. Springer International Publishing, Switzerland. (pp. 105–130).

McKinnon, Innisfree, & Hiner, Colleen. (2016). Does the region still have relevance? (Re)considering 'regional' political ecology. *Journal of Political Ecology, 23*, 115–122.

McLain, Rebecca J., Hurley, Patrick T., Emery, Marla R., & Poe, Melissa R. (2013). Gathering 'wild' food in the city: Rethinking the role of foraging in urban ecosystem planning and management. *Local Environment, 19*(2), 220–240. Routledge.

McLain, Rebecca, Poe, Melissa, Hurley, Patrick T., Lecompte-Mastenbrook, Joyce, & Emery, Marla R. (2012). Producing edible landscapes in seattle's urban forest. *Urban Forestry & Urban Greening, 11*(2), 187–194.

Mercer, Claire. (2016). Landscapes of extended ruralisation: Postcolonial Suburbs in Dar Es Salaam, Tanzania. *Transactions of the Institute of British Geographers, 42(1), 72-83*.

Momsen, Janet Henshall. (1996). Agriculture in the Sierra. In Don C. Erman (Ed.), *Sierra nevada ecosystem project: Final report to congress* (pp. 497–528).

Monstadt, Jochen. (2009). Conceptualizing the political ecology of urban infrastructures: Insights from technology and urban studies. *Environment and Planning A, 41*(8), 1924–1942.

Nelson, Lise, & Nelson, Peter B. (2010). The global rural: Gentrification and linked migration in the rural USA. *Progress in Human Geography, 35*(4), 441–459.

Newell, J. P., & Cousins, J. J. (2014). The boundaries of urban metabolism: Towards a political-industrial ecology. *Progress in Human Geography, 39*(6), 702–728.

Nik., Heynen, Kaika, Maria, & Swyngedouw, Erik. (2006). *The nature of cities: Urban political ecology and the politics of urban metabolism* (1st ed.). New York, NY: Routledge.

Olson, Jeffrey L., & Munroe, Darla K. (2012). Natural amenities and rural development in new urban-rural spaces. *Regional Science Policy & Practice, 4*(4), 355–371.

Parés, Marc, March, Hug, & David, Saurí. (2013). Atlantic gardens in mediterranean climates: Understanding the production of suburban natures in Barcelona. *International Journal of Urban and Regional Research, 37*(1), 328–347.

Peet, Richard, & Watts, Michael (Eds.). (1996). *Liberation ecologies*. New York, NY: Routledge.

Poe, Melissa R., LeCompte, Joyce, McLain, Rebecca, & Hurley, Patrick. (2014). Urban foraging and the relational ecologies of belonging. *Social & Cultural Geography, 15*(8), 901–919. Routledge.

Poe, Melissa R., McLain, Rebecca J., Emery, Marla, & Hurley, Patrick T. (2013). Urban forest justice and the rights to wild foods, medicines, and materials in the city. *Human Ecology, 41*(3), 409–422.

Quastel, Noah. (2009). Political ecologies of gentrification. *Urban Geography, 30*(7), 694–725.

Reed, Maureen G. (2007). Uneven environmental management: A canadian comparative political ecology. *Environment and Planning A, 39*(2), 320–338.

Robbins, P, & Sharp, JT. (2003). Producing and consuming chemicals: the moral economy of the american lawn. *Economic Geography, 79*(4), 425-51. doi:10.1111/ecge.2003.79.issue-4

Robbins, Paul. (2002). Letter to the editor: Obstacles to a first world political ecology? Looking near without looking up. *Environment and Planning A, Environment and Planning A, 34*(8), 1509-1513. Retrieved from http://ideas.repec.org/a/pio/envira/v34y2002i8p1509-1513.html

Robbins, Paul. (2007). *Lawn people: How grasses, weeds, and chemicals make us who we are.* Philadelphia, PA: Temple University Press.

Robbins, Paul. (2012). *Political Ecology : A Critical Introduction.* Hoboken, NJ: Wiley-Blackwell.

Robbins, Paul, & Bishop, Kristina Monroe. (2008). There and back again: Epiphany, disillusionment, and rediscovery in political ecology. *Geoforum, 39*(2), 747-755.

Robbins, Paul, Martin, Stephen, & Gilbertz, Susan. (2011). Developing the commons: The contradictions of growth in Exurban Montana. *The Professional Geographer, 64*(3) 317-331. Taylor & Francis.

Sandoval, Gerardo Francisco, & Maldonado, Marta Maria. (2012). Latino urbanism revisited: Placemaking in new gateways and the urban-rural interface. *Journal of Urbanism: International Research on Placemaking and Urban Sustainability, 5*(2-3), 193-218. Routledge.

Sayre, Nathan Freeman. (2002). *Ranching, endangered species, and urbanization in the Southwest : Species of capital.* Tucson, AZ: University of Arizona Press.

Scott, Alister J., Shorten, James, Owen, Rosalind, & Owen, Iwan. (2009). What kind of countryside do the public want: Community visions from wales UK? *GeoJournal, 76*(4), 417-436.

Sieverts, Thomas. (2003). *Cities without Cities : An Interpretation of the Zwischenstadt.* London, New York: Spon Press.

Silver, Jonathan. (2015). Disrupted infrastructures: An urban political ecology of interrupted electricity in Accra. *International Journal of Urban and Regional Research, 39*(5), 984-1003.

Smith, Michael D., & Krannich, Richard S. (2000). "Culture clash" revisited: Newcomer and longer-term residents' attitudes toward land use, development, and environmental issues in rural communities in the rocky mountain west*. *Rural Sociology, 65*(3), 396-421.

Smith, Neil. (2008). *Uneven development: Nature, capital, and the production of space.* Athens, GA: University of Georgia Press.

Swyngedouw, Erik. (1997). Power, nature, and the city. The conquest of water and the political ecology of urbanization in Guayaquil, Ecuador: 1880-1990. *Environment and Planning A, 29* (2), 311-332. Pion Ltd.

Swyngedouw, Erik. (1999). Modernity and hybridity: Nature, regeneracionismo, and the production of the Spanish waterscape. *Annals of the Association of American Geographers, 89*(3), 443.

Swyngedouw, Erik. (2004). *Social Power and the urbanization of water: Flows of power.* USA: Oxford University Press.

Taylor, Laura E, & Hurley, Patrick T. (2016). The broad contours of Exurban landscape change. In Laura E. Taylor & Patrick T. Hurley (Eds.), *A comparative political ecology of exurbia: Planning, environmental management, and landscape change.* Cham, Switzerland: Springer International Publishing.

Theobald, David M. (2005). Landscape patterns of Exurban growth in the USA from 1980 to 2020. *Ecology And Society, 10*(1), 32.

Todd, Nesbitt J., & Weiner, Daniel. (2001). Conflicting environmental imaginaries and the politics of nature in central Appalachia. *Geoforum, 32*(3), 333-349. Elsevier.

Trubek, Amy B., & Bowen, Sarah. (2008). Creating the taste of place in the United States: Can we learn from the French? *GeoJournal, 73*(1), 23-30.

Turner, Billie L., & Robbins, Paul. (2008). Land-change science and political ecology: Similarities, differences, and implications for sustainability science. *Annual Review of Environment and Resources, 33*(1), 295-316. Annual Reviews.

Unwin, Tim. (2012). Terroir: At the heart of geography. In Percy H. Dougherty (Ed.) *The geography of wine* (pp. 37-48). Dordrecht: Springer Netherlands.

Walker, Peter A., & Hurley, Patrick T. (2011). *Planning paradise: Politics and visioning of land use in oregon.* Tucson, AZ: University of Arizona Press.

Walker, Peter, & Fortmann, Loise. (2003). Whose landscape? A political ecology of the 'Exurban' Sierra. *Cultural Geographies, 10*(4), 469–491.

Walker, Richard. (2015). Building a better theory of the urban: A response to 'Towards a new epistemology of the urban?'. *City, 19*(2–3), 183–191. Routledge.

Watts, Michael, & Peet, Richard. (2004). Liberating political ecology. In Richard Peet, Michael Watts (Eds.) *Liberation ecologies: Environment, development, social movements* (pp. 3–43). Routledge.

Woods, Michael. (2011). Rural geography III: Rural futures and the future of rural geography. *Progress in Human Geography, 36*(1), 125–134.

Index

Note: **Bold** page numbers refer to tables; *italic* page numbers refer to figures and page numbers followed by "n" denote endnotes.

Abrams, J. B. 103
Adger, W. 51
Agrawal, A. W. 28
air pollution 17
alternative fuels: built environment system 20–24, **21–22**; costs and subsidies 27; promising research directions 27–32
alternative-fuel vehicles (AFV) 27
American Disability Act 75
Angelo, H. 93
Ashmore, P. 3
ASIF framework 15

Behrens, W. 15
Beilin, R. 53, 58n4
Benedict, M. A. 53, 65
Berke, P. R. 71
big science 5
biophysical environment 3–4
Blaikie, P. 88
Bolund, P. 65
Boone, C. 73
Brenner, N. 88
Brown, B. 27
Brown, K. 48
The Brundtland Commission 13
Bruntland Report 46
built environment system 20–24, **21–22**; alternative modes 23; land-use strategies 23–24; network configuration 24; pricing 23

Cadenasso, M. 67
Campbell's Planner's Triangle 14
CAVs *see* connected and autonomous vehicles (CAVs)
C40 Cities 32
Chelleri, L. 52
Circella, G. 27

climate change 17
Concrete and Clay: Reworking Nature in New York (Gandy, M.) 73
connected and autonomous vehicles (CAVs) 30–31
Conroy, M. M. 71
Cooper, M. 48
Corporate Average Fuel Economy (CAFE) standards 19
Coupled Natural Human Systems Program 5
Cousins, J. J. 91

David, S. 93
Derickson, K. D. 47, 48
Dill, J. 28
Dodson, B. 3
Dunn, A. D. 73, 75

economic theory 18
empirical research 53–57; green infrastructure scenarios comparison 55–57, *56–57*; green space 55; stormwater management benefits 55
equitable transportation systems 18
e-shopping 28

five Ws 50–53, **54**
Fortmann, L. 89
Future Earth 5

Gallagher, R. 28
Gandy, M. 91
geographic information systems (GIS) 70
Giuliano, G. 26
global cities 99
Global Land Programme (GLP) 5
Gosnell, H. 103
Grand Rapids's plan 71
greenhouse gas (GHG) emissions 19

green infrastructure theory: definition 65; distributional equity 73–77, **74**, *76–77*; equity in process 77–79; mapping distribution 69–71; multiple benefits 74, **74**; natural ecosystem values and functions 66; place and locale 71–72; place-based understanding of 79; private green space 70; procedural equity 72–73; scale relevant to 68–69; socio-ecological analysis 73; theoretical and practical challenges 79; unified conceptualization of 79
green space connectivity 69
green space inequity 73
green sprawl process 103
Green Visions Plan 55
Gunderson, L. 52
Güneralp, B. 67
Guo, Z. 28
Gustafson, S. 93

Hamstead,, Z. A. 67
Handy, S. L. 27
Harden, C. P. 104
Harvey, D. 91
Heynen, N. 73
high-speed rail (HSR) 29
hobby farms 101
Holling, C. S. 45, 52
Huang, G. 67
human-biophysical dynamics 2
human-environment geography 2–3; land systems science approaches 2; urbanization science approach 2
Hunhammar, S. 65
Hurley, P. K. 88
Hutrya, L. R. 67
hydraulic fracturing technology 17

Ignatieva, M. 66
integrated political ecology approach 92–94
integrated (ex) urban political ecology 88–89
intelligent transportation systems (ITS) 20
Intergovernmental Panel on Climate Change (IPCC) 5
International Geosphere-Biosphere Programme (IGBP) 5
international resilience policy 58n1
ITS *see* intelligent transportation systems (ITS)

Joseph, J. 47
Journal of the American Planning Association 25

Kaika, M. 93
Keeley, M. 6, 7
Keil, R. 93
Kelman, I. 44
Kremer, P. 67

land systems science approaches: foundational work in 2
Lee, B. 28
Lee, E. 28
Lefebvre's theory 87
The Limits to Growth 15
Long Term Ecological Research (LTER) programs 66
Los Angeles Council District scale 57, *57*
Lovejoy, K. 27
Luke, T. W. 99

MacKinnon, D. 47, 48
March, H. 93
Marxist theory 91
McDonnell, M. 67
McKinnon, I. 6–8
McMahon, E. T. 53, 65
McPhearson, T. 67
Meadows, D. H. 15
Meadows, H. 15
Meerow, S. 6–8, 49
Mell, I. C. 66
Meurk, C. 66
Miller, F. 46
Mokhtarian, P. L. 27

National Climate Assessment 31, *31*
National Science Foundation (NSF) 5
Newell, J. P. 6–8, 91
new localism 64
new theories 7–8
non renewable fossil fuels 13
non-timber forest products 90
non-urban systems 2

Ocean Extended Vision Plan 59n5
oil prices 15, **16**, 17
oil reserves 15, **16**, 17
oil supply 15, **16**, 17
Olazabal, M. 52
One Million Trees program 70

Parés, M. 93
Paris Agreement's 19
Parkany, G. 28
Pearsall, H. 72
Pickett, S. T. 67
Pierce, J. 72
The Planner's Triangle 25, *25*
political ecology 3
political-economic system 91
promising research directions 27–32; adaptation to climate change 31–32; alternative-fuel vehicle adoption 30; alternative-fuel vehicle station 30; attitudes affect 28; connected and

autonomous vehicles 30–31; cultural and developmental contexts 29; disadvantaged neighborhoods 29; high-speed rail 29–20; hyperloop stations 29–20; include preferences 28; natural experiments 27; non-electric AFVs 30; past experiences 28; policy and investment packages 28; policy experiments 27; psychological well-being effect 30; social equity 29; understanding heterogeneity 28; urban freight 28
public open space 66

Randers, D. L. 15
Redman, C. L. 46
resilience theory 57–58
road safety 18
Robbins, P. 6, 93, 104

Salon, D. 23, 27
Salt, D. 52
Schmid, C. 88
Sciara, G.-C. 27
SES *see* social-ecological systems (SES)
Seto, K. C. 67
Sloan, A. 66
small-scale green infrastructure 65
smart cities 46
Smith, N. 91
social-ecological systems 6, 67, 69
social-ecological systems (SES) 44, 45
social theory 1
spatial theory 67
Spears, S. 27
Stewart, G. H. 66
stormwater management systems 69
Sultana, S. 6, 7
sustainable transportation technology: alternative fuels 19–20; alternative stations 19–20; alternative vehicles 19–20; cleaner petroleum-powered cars 19; intelligent transportation systems 20; techno-centric approaches 19
Sustainable Transportation Technology approach 14
Swyngedouw, E. 93

Taylor, L. E. 88
traffic congestion 17
transit-oriented development (TOD) 25–26
transport affordability 18
transport equity 18–19
transport revolution 32
tree planting programs 78
Turner, B. L. 104

urban age 86
urban food 74

urban freight 28
Urban Geography 3, 8
urban green infrastructure 54
urban green space 66
Urbanization and Global Environmental Change (UGEC) 5
urbanization science 1, 5
urban metabolism 91
urban open space 66
urban political ecology 88; extended urbanization 88–89; exurban political ecology 89–92; Gustafson's conceptualization 93; integrated approach 92–94; integrated (ex) urban political ecology 88–89; non-timber forest products 90; political ecology approaches 87; shifting dynamics 94–102; urban sustainability 102–103; *vs.* exurban political ecology approaches 92, *92*
urban resilience: adaptation scholarship 46–47; boundary object 43, 49–50; concept of 43, *44*; empirical contexts 53–55; empirical research 53–57; five Ws 50–53, **54**; green infrastructure requires 54; politics of *48*, 48–53; sustainability 46–47; theoretical critiques 47–48; vulnerability scholarship 46–47
urban resilience theory 51
urban rural interfaces 92–94; jackson county oregon 99–102; Sierra Nevada mountains 97–99; "Stone Hill" area 94–97
Urban Science 5
urban sustainability: an urbanizing world 102–103; political ecology 103–105
urban sustainability scholarship: biophysical environment in 3–4; entrenching divides 4–6; human-environment geography 2–3; integrative geographical research 6; land system science 2; macro-scale issues 6; new theories 7–8; potential strength 7; systems thinking 6; techno-engineering problem 7
urban system 49; conceptual model 49, *50*
urban transport systems: air pollution 17; carrying capacity 15, **16**; climate change 17; health promotes 19; material throughput 15, **16**; oil prices 15, **16**, 17; oil reserves 15, **16**, 17; oil supply 15, **16**, 17; physical inactivity 19; road safety 18; sustainable transportation technology approach 14, *15*, 19–20; traffic congestion 17; transport affordability 18; transport equity 18–19; win-win-win solutions 24–27, *25*

Vale, L. J. 49, 51, 53, 58n3
Van Acker, V. 28
Van Wee, B. 28
Vision Zero 18
Viveiros, P. 28

Wachsmuth, D. 93
Wagenaar, H. 52
Walker, B. 48, 52
Weber, T. 66
Weichselgartner, J. 44
Wheeler, J. 3
Wilkinson, C. 52, 53, 58n4
win-win-win solutions 24–27, *25*; affordable housing provision 26; alt-fuel vehicle costs 27; alt-fuel vehicle subsidies 27; gentrification 25–26; infill development 26; transit-oriented development 25–26; transit's multiple purposes 26
Witlox, F. 28
Wolf, J. 66
Wu, J. 52
Wu, T. 52

Ye, L. 27
Young, D. 93

Zhou, W. 67